ADVANCES IN ENGINEERING RESEARCH

ADVANCES IN ENGINEERING RESEARCH

VOLUME 30

ADVANCES IN ENGINEERING RESEARCH

Additional books and e-books in this series can be found on Nova's
website under the Series tab.

ADVANCES IN ENGINEERING RESEARCH

VOLUME 30

VICTORIA M. PETROVA
EDITOR

Copyright © 2019 by Nova Science Publishers, Inc.

All rights reserved. No part of this book may be reproduced, stored in a retrieval system or transmitted in any form or by any means: electronic, electrostatic, magnetic, tape, mechanical photocopying, recording or otherwise without the written permission of the Publisher.

We have partnered with Copyright Clearance Center to make it easy for you to obtain permissions to reuse content from this publication. Simply navigate to this publication's page on Nova's website and locate the "Get Permission" button below the title description. This button is linked directly to the title's permission page on copyright.com. Alternatively, you can visit copyright.com and search by title, ISBN, or ISSN.

For further questions about using the service on copyright.com, please contact:
Copyright Clearance Center
Phone: +1-(978) 750-8400   Fax: +1-(978) 750-4470   E-mail: info@copyright.com.

NOTICE TO THE READER

The Publisher has taken reasonable care in the preparation of this book, but makes no expressed or implied warranty of any kind and assumes no responsibility for any errors or omissions. No liability is assumed for incidental or consequential damages in connection with or arising out of information contained in this book. The Publisher shall not be liable for any special, consequential, or exemplary damages resulting, in whole or in part, from the readers' use of, or reliance upon, this material. Any parts of this book based on government reports are so indicated and copyright is claimed for those parts to the extent applicable to compilations of such works.

Independent verification should be sought for any data, advice or recommendations contained in this book. In addition, no responsibility is assumed by the Publisher for any injury and/or damage to persons or property arising from any methods, products, instructions, ideas or otherwise contained in this publication.

This publication is designed to provide accurate and authoritative information with regard to the subject matter covered herein. It is sold with the clear understanding that the Publisher is not engaged in rendering legal or any other professional services. If legal or any other expert assistance is required, the services of a competent person should be sought. FROM A DECLARATION OF PARTICIPANTS JOINTLY ADOPTED BY A COMMITTEE OF THE AMERICAN BAR ASSOCIATION AND A COMMITTEE OF PUBLISHERS.

Additional color graphics may be available in the e-book version of this book.

Library of Congress Cataloging-in-Publication Data

ISBN: 978-1-53616-092-5
ISSN: 2163-3932

Published by Nova Science Publishers, Inc. † *New York*

CONTENTS

Preface		**vii**
Chapter 1	Representation of Gaussian Beams Using Rays *Paul D. Colbourne*	**1**
Chapter 2	Novel I/O-LDC Control Based on ANFIS for Variable Speed Wind-Turbine System *Fayssal Amrane and Azeddine Chaiba*	**47**
Chapter 3	Corrosion Effects on Steel Reinforcement *Ch. Apostolopoulos and Arg. Drakakaki*	**111**
Chapter 4	Development of an Online Inspection System for Computer Numerical Control Cutting Lathe Tool Inserts Using Eye-in-Hand Machine Vision *Wei-Heng Sun and Syh-Shiuh Yeh*	**143**
Chapter 5	Type-2 Fuzzy Logic Control for Quadrotor *Gherouat Oussama, Fayssal Amrane* *and Abdelouahab Hassam*	**167**
Chapter 6	Using 2D Spatial Filtering Techniques to Simulate Surface Single Fibre Action Potential *Noureddine Messaoudi, Raïs El'hadi Bekka* *and Samia Belkacem*	**197**
Contents of Earlier Volumes		**215**
Index		**219**

PREFACE

Advances in Engineering Research. Volume 30 opens by presenting methods in which rays can be used to calculate the properties of a propagating generally astigmatic Gaussian beam at any point in an optical system, optimize the system to obtain the desired Gaussian beam focus and minimize aberrations, analyze stray light, or calculate fiber coupling efficiency.

The authors also present an enhanced input/output-linearizing and decoupling control in variable speed for a wind energy conversion system, which is based on an adaptive neuro-fuzzy inference system, a combination of two soft-computing methods of artificial neural networks and fuzzy logic.

Reinforced concrete, one of the world's most common building materials, is explored. Its utility and versatility are achieved through the combination of concrete and steel, however, the deterioration of such structures is a frequent phenomenon due to environmental factors.

The authors go on to address the need for online insert inspection during computer numerical control lathe cutting processes through the development of a system based on eye-in-hand machine vision that incorporates a manipulator. This system can perform online inspections for detection and classification of fractures, built-up edges, chipping, and flank wear in the inserts of external turning tools.

Next, type-2 fuzzy logic control is proposed in order to control the altitude and attitude of an unmanned quadrotor. Simulation results suggest that the proposed control provides improved dynamic responses and perfect decoupled control in steady and transient states.

In closing, an analytical approach called the 2D spatial filtering technique was used to simulate the surface single fibre action potential signal. The study demonstrates that the surface single fibre action potential generated by an impulsive source is the outcome of the spatial phenomena.

Chapter 1 - Methods are presented of using rays to calculate the properties of a propagating generally astigmatic Gaussian beam at any point in an optical system, and to optimize a generally astigmatic optical system to obtain the desired Gaussian beam focus and minimize aberrations. The optimization method requires very little computation beyond that of a conventional ray optimization, and requires no explicit calculation of the properties of the propagating Gaussian beam. Unlike previous methods, the calculation of beam parameters does not require matrix calculations or the introduction of non-physical concepts such as imaginary rays. A ray set can be generated which has a ray density proportional to the Gaussian field amplitude or intensity at all points in space, which can be used for Monte Carlo analyses of stray light. Such a ray set can also be used to compute the output field amplitude and the coupling efficiency into a fiber or waveguide mode. A simplified method is presented for computing the coupling efficiency when the output mode has a Gaussian profile.

Chapter 2 - This chapter presents an enhanced Input/Output-Linearizing and Decoupling Control (I/OLDC) in variable speed for Wind Energy Conversion System (WECS), which is based on ANFIS (Adaptive Neuro-Fuzzy Inference System). ANFIS is a combination of two soft-computing methods of ANN (Artificial Neural Network) and fuzzy logic. Fuzzy logic has the ability to change the qualitative aspects of human knowledge and insights into the process of precise quantitative analysis. The improved I/OLDC is applied in variable speed for WECS using Doubly Fed Induction Generator (DFIG) fed by back-to-back two-phase voltage source inverter (VSI). The proposed control technique is used in order to control the stator active and reactive powers of DFIG by the means of MPPT (Maximum Power Point Tracking) strategy. By using the feedback linearization, the control algorithm is established. In order to improve the tracking of stator active and reactive power references, ANFIS is used after the comparison between measured and reference of stator active and reactive powers respectively. The proposed control is used to overcome the drawbacks of the classical control such as PI controllers in terms of; overshoot, response time, power error and the reference tracking. Finally, simulation results demonstrate that the proposed control using ANFIS provides improved dynamic responses and perfect decoupled control of the wind turbine has driven DFIG with high performances (good reference tracking, short response time and neglected power error) in steady and transient states.

Chapter 3 - Reinforced concrete is one of the world's most common building materials. Its utility and versatility are achieved through the combination of the best features of concrete and steel. Thanks to concrete a great resistance to compression can be recorded and thanks to steel high ductility and consequently significant resistance to tension can be developed. However, deterioration of such structures, due to ageing, which is owed to environmental factors, is a frequent phenomenon.

Among the main ageing factors though, is corrosion due to chlorides penetration. Mechanical degradation of steel, concrete spalling or even premature failure are some of the main consequences of diffusion of chlorides, which primarily initiate from areas where poorly processed concrete or inadequate concrete thickness can be found. Of course, permeability of concrete, combined with existing surface defects, or even subsistent corrosion damage detected in steel, is responsible for harsher damage, which in long-term can be proved destructive and liable to cause unexpected consequences or situations. Additional factors, which determine significantly the severity of damage, besides the environmental parameters, are the length of the reinforcement, which is exposed to the aggressive conditions -each time- in function of the length of the adjacent areas, which are still protected by concrete, due to the "differential aeration corrosion" phenomenon. Therefore, structural elements with exposed areas of different length, on their reinforcement, are expected to demonstrate uniform level of corrosion damage, which consequently results in differentiated mechanical degradation and unavoidably dissimilar performance against imposed loads.

Analogous parameters and remarks raise several issues, and trigger the scientific community's interest on the laboratory investigation of such topics. Two of the most popular methods, which are used for the laboratory simulation and reproduction of the corrosive conditions, are "salt spray chamber" and "impressed current density" technique. Salt spray chamber tests are carried out in accordance to ASTM B117 specification; however, there are no similar specifications available for impressed current density technique.

Chapter 4 - To address the online insert inspection need during computer numerical control (CNC) lathe cutting processes, the authors developed a system based on eye-in-hand machine vision that incorporates a manipulator. This system can perform online inspections for detection and classification of fractures, built-up edges (BUEs), chipping, and flank wear in the inserts of external turning tools. To circumvent the influence of the machining environment on the acquisition of insert images, a machine vision module equipped with a surrounding light source and a fill-light for insert-tips was designed, which could be installed on a manipulator. This manipulator allows the machine vision module to reach the lathe turret position and capture clear and detailed images of the lathe's inserts. Using this strategy, subsequent image processing procedures and insert status judgment are significantly more straightforward. In the process of capturing insert images, the intensity of the surrounding light source and insert-tip fill-light varied so that the surface features and tip features in the insert images could be readily identified. Insert inspection and classification were performed according to the following procedure: 1) construction of the insert's outer profile, 2) capture of the insert's status region, and 3) determination and calculation of the insert's wear region. To obtain images of the insert's wear region, insert images were trimmed according to the vertical flank line, horizontal blade line, and vertical blade line of the insert, which were, in turn, constructed from its outer profile features. The occurrence of flank or chipping wear was determined according to gray-level histograms. The images of the wear region were

Preface   xi

used to calculate the amount of wear for evaluating and assessing whether an insert had normal or excessive levels of wear. In the present study, insert inspection experiments were performed to establish that the developed machine vision module and computational methods could detect insert fractures, BUEs, chipping, and flank wear. When the positioning offset of the manipulator-operated machine vision module was lower than 0.5 mm, the chipping rate detection error was less than 8%, and the detection error of the amount of wear was within 2 pixels. These results establish the viability and efficacy of the developed online eye-in-hand lathe inspection system.

Chapter 5 - In this chapter Type-2 Fuzzy logic Control (T2-FLC) is proposed in order to control the altitude and attitude of an unmanned quadrotor. The mathematical model of six degree of freedom (6DOF) aerial vehicle is nonlinear and highly-coupling, and under-actuated; all these drawbacks make the system unstable. For the whole-system; the partials six controls were distributed and select the membership's functions carefully due of the impact on each other. However, the memberships functions (MFs) are the heart of any fuzzy logic system the fuzzy type-2 by nature can handle better uncertainty and solve the strong coupling and trajectory tracking problem. In order to assess the fuzzy type-2 controller performance, it has been compared with a Sliding Mode Controller (SMC) for two popular different trajectories straight lines and Helical. The proposed control based on T2-FLC is used to overcome the drawbacks of the classical control based on SMC in terms of; overshoot, tracking accuracy and control efforts. Finally, simulation results prove that the proposed control provides an improved dynamic responses and perfect decoupled control in steady and transient states.

Chapter 6 - Surface single fibre action potential (SFAP) is the main component of the surface electromyographic (EMG) signal. In this chapter, an analytical approach called "two-dimensional (2D) spatial filtering techniques" was used to simulate the surface SFAP signal. The volume conductor and the detection system were considered as 2D spatial filter, its input signal was the current density source and its output signal was the potential detected on the skin surface. The volume conductor was considered as a planar, space-invariant, multilayer (non-homogeneous) and anisotropic medium constituted by muscle, fat and skin tissues. The detection system was obtained by combining the spatial filter, size and shape of the electrodes. One (1D) and two (2D) spatial filters were investigated. Rectangular, circular, elliptical and annular shaped electrodes were used to detect the SFAP signal. Two current density (impulsive and analytical) sources were used for the generation of surface SFAP signal and a comparison study was made. The authors showed that the surface SFAP generated by an impulsive source is the outcome of the spatial phenomena and the one generated by an analytical source is due to the diameter and to the intracellular conductivity of the fibre.

In: Advances in Engineering Research
Editor: Victoria M. Petrova

ISBN: 978-1-53616-092-5
© 2019 Nova Science Publishers, Inc.

Chapter 1

REPRESENTATION OF GAUSSIAN BEAMS USING RAYS

Paul D. Colbourne[*]
Lumentum, Ottawa, Canada

ABSTRACT

Methods are presented of using rays to calculate the properties of a propagating generally astigmatic Gaussian beam at any point in an optical system, and to optimize a generally astigmatic optical system to obtain the desired Gaussian beam focus and minimize aberrations. The optimization method requires very little computation beyond that of a conventional ray optimization, and requires no explicit calculation of the properties of the propagating Gaussian beam. Unlike previous methods, the calculation of beam parameters does not require matrix calculations or the introduction of non-physical concepts such as imaginary rays. A ray set can be generated which has a ray density proportional to the Gaussian field amplitude or intensity at all points in space, which can be used for Monte Carlo analyses of stray light. Such a ray set can also be used to compute the output field amplitude and the coupling efficiency into a fiber or waveguide mode. A simplified method is presented for

---

[*] Corresponding Author's Email: paul.colbourne@lumentum.com.

Paul D. Colbourne

computing the coupling efficiency when the output mode has a Gaussian profile.

Keywords: Gaussian beam, ray, optics, optimization, general astigmatism

1. INTRODUCTION

Although several authors have shown that propagating rays can represent propagating Gaussian beams [1-8], these techniques are not well known among optical designers. Steier [1] showed that a ray packet consisting of a family of an infinite number of rays propagates in a manner equivalent to the propagation of a Gaussian beam. Arnaud [2] showed that Gaussian beam propagation can be described by the propagation of "complex rays" (a complex ray consisting of a "real" ray and an "imaginary" ray, essentially two rays), and he also showed that skew rays follow the envelope of a propagating circular Gaussian beam. Herloski et al. [3] have shown that by propagating two carefully selected rays through a simply astigmatic optical system (separable into x and y components), simple calculations allow the determination of the properties of a propagating Gaussian beam. Because of their simplicity, the techniques of Herloski et al. are commonly used by optical design software to compute Gaussian beam parameters, but the results are not correct for the case of general astigmatism, where the optics need not be aligned with the x-y axes. Greynolds [4], Arnaud [5], and Lü et al. [6] have shown that by propagating four rays (two "real" rays and two "imaginary" rays), complex matrix calculations can be used to determine Gaussian beam properties in the case of general astigmatism. Colbourne [7] showed that by using four rays, simple calculations analogous to those of Herloski et al. [3] can provide the properties of a propagating Gaussian beam in the case of general astigmatism, without need for matrix calculations or non-physical concepts such as complex rays. The techniques of Colbourne extend the simple approach of Herloski et al. to generally astigmatic systems, thus removing any constraints on the orientations of optical elements.

Calculation of propagating Gaussian beam parameters is useful, but also of interest would be to use rays to optimize an optical system which uses Gaussian beams. Available techniques for this are limited and have various disadvantages. The formulae of Herloski et al. [3] can be used to obtain the correct focus, but these formulae only work for the paraxial case, and they do not apply to systems with general astigmatism. This is a more serious restriction than may be initially thought, because even if an optical system does not contain any intentionally rotated elements, a spherical lens can induce general astigmatism if the beam is incident off-axis, or elements may become rotated during a tolerancing exercise. Pakhomov and Tsibulya [8] describe a ray-based optimization method for use with Gaussian beams which can correct aberrations, but again their method is limited to the case of simple astigmatism. Using conventional rays to optimize aberrations may result in a system that is not properly optimized, even if the resulting focus errors are not important, because the rays used do not sample the actual lens apertures which the Gaussian beam sees. For example, at a particular surface the conventional rays may focus to a single point, whereas the actual beam can occupy a significant region of the optic. Beam propagation methods can be used for Gaussian beams, but are computationally intensive, may not work through all surface types, and tend not to converge well because calculation "noise" can cause the optimization algorithms to get stuck in local minima. What is desired is a computationally efficient ray-based optimization method applicable to generally astigmatic systems. Colbourne [7, 9] proposed and demonstrated a ray-based optimization method compatible with general astigmatism and capable of correcting aberrations which makes use of launched skew rays instead of the conventional method of rays diverging from a point. Incorporating these methods into ray-tracing software should enable faster, more accurate design of optical systems which use Gaussian beams.

This chapter describes the techniques presented by Colbourne [7] with some updates and extensions, such as the use of direction cosines instead of ray slopes in the ray definitions, which provides more accurate results for beams with high divergence. Section 2 gives the basics of using skew rays to represent a Gaussian beam. Also presented is a method of

4 Paul D. Colbourne

producing a set of rays with a ray density proportional to the field amplitude or intensity of a propagating Gaussian beam. Such a set of rays can be used to analyze stray light in optical systems employing Gaussian beams. Section 3 gives methods for calculating beam properties using rays. Section 4 describes methods for optimizing optical systems using rays, including correction of aberrations. Section 5 outlines a method for using skew rays to calculate an output field profile and fiber coupling efficiency.

2. SKEW RAY REPRESENTATION OF GAUSSIAN BEAMS

Circular Beams

I will first review the skew ray representations of circular and simply astigmatic Gaussian beams, then I will extend the representation to generally astigmatic systems. As Arnaud explained [1] (see his Figure 1), a set of skew rays will follow the envelope of a propagating circular Gaussian beam. This construction is of more than theoretical interest, it is useful for visualizing the propagation of a Gaussian beam through an optical system. Starting at the input beam waist, the set of skew rays is defined by

$$x = \omega_0 \cos(\alpha)$$
$$y = \omega_0 \sin(\alpha)$$
$$L = -\theta_0 \sin(\alpha)$$
$$M = \theta_0 \cos(\alpha) \tag{1}$$

where L and M are direction cosines, that is, L is the cosine of the angle between the ray and the x axis and M is the cosine of the angle between the ray and the y axis, ω_0 is the waist radius, $\theta_0 = \lambda/(n \pi \omega_0)$ is the divergence angle of the beam, λ is the wavelength, and n is the refractive index at the input. The third direction cosine $N = \sqrt{1 - L^2 - M^2}$ is needed for entering the ray direction into ray-tracing software. Note that Arnaud and others (including my previous work [7]) use ray slopes $l = dx/dz$ and $m = dy/dz$ in

Equation (1), not the direction cosines L and M, which makes negligible difference for small beam divergence, but at higher divergence one can make at least two arguments in support of using direction cosines. First, if ray slopes are used, arbitrarily small values of ω_0 could be defined, whereas in fact there is a lower limit of about $\lambda/2$ imposed by diffraction. Second, if we consider the case of a Gaussian beam waist located at a planar boundary between different refractive indices, Snell's law of refraction suggests that L and M should be used in Equation (1), not l and m. At the end of Section 5, an additional argument in support of using L and M is made based on better field calculations.

If an infinite number of rays are generated, with all values of α from 0 to 2π, the resulting ray bundle will coincide with the $1/e^2$ intensity boundary of the propagating Gaussian beam at all points through the optical system, provided the beam remains circular and there are no aberrations. This is equivalent to Steier's ray packet representation of a Gaussian beam [1]. Propagating a subset of rays, say 16 rays with α equally spaced between 0 and 2π, allows easy visualization of Gaussian beams on layout diagrams of ray-tracing software; an example is shown in Figure 1.

Herloski et al.'s construction [3] of a "waist ray" (originating a distance ω_0 from the z axis, parallel to the z axis) and a "divergence ray" (originating at the origin at an angle corresponding to the far-field divergence angle θ_0 of the Gaussian beam) can be interpreted as a projection of two skew rays with $\alpha = 0$ and $\alpha = \pi/2$ in Equation (1) onto the x-z plane. It is worth noting that any two skew rays with α differing by $\pi/2$ may be used as the launched rays, and all of Herloski et al.'s results will still be valid. The choice of rays to launch is not limited to the waist ray and divergence ray defined by Herloski et al.

Figure 1. An example of the use of skew rays to visualize the propagation of Gaussian beams in optical systems.

6 *Paul D. Colbourne*

Elliptical Beams

If the optical system is simply astigmatic, that is, the input elliptical beam is aligned with the x-y axes and all cylindrical or astigmatic lenses are aligned with the x-y axes, one can construct a waist ray and divergence ray for the x axis and a waist and divergence ray for the y axis and compute the x and y components of the propagating Gaussian beam independently. One can realize these two waist rays and two divergence rays as projections of the same two skew rays onto the x-z plane and the y-z plane, thus by propagating two skew rays one can compute the x and y parameters of the propagating simply astigmatic Gaussian beam. A skew ray bundle defined by

$$x = \omega_{0x}\cos(\alpha)$$
$$y = \omega_{0y}\sin(\alpha)$$
$$L = -\theta_{0x}\sin(\alpha)$$
$$M = \theta_{0y}\cos(\alpha) \tag{2}$$

would create the same waist rays and divergence rays defined by Herloski when rays corresponding to $\alpha = 0$ and $\alpha = \pi/2$ are projected onto the x-z and y-z planes. Again, any two skew rays with α differing by $\pi/2$ may be launched and all of Herloski's analysis results will still be correct.

Generally Astigmatic Beams

While Herloski's analysis works for elliptical beams in simply astigmatic optical systems, the skew ray bundle defined by Equation (2) does not follow the envelope of an elliptical beam. This is illustrated in Figure 2, which shows the evolution of a skew ray bundle representing an elliptical beam according to Equation (2). The skew ray bundle remains elliptical in shape but the major axis rotates as the bundle propagates, however we know that a simple elliptical beam remains aligned with the x-y axes as it propagates. The skew ray bundle does however follow the profile of the elliptical beam if viewed only along the x or y direction, as it must according to Herloski's analysis.

The skew ray bundle may be defined with a right handed skew (that is, the rays twist as a right-handed screw as the rays propagate) or with a left handed skew, either sense would be equivalent to Herloski et al.'s construction. The rays generated by Equation (2) are right skew rays; generating rays with a left handed skew involves using the opposite sign for either x and y, L and M, x and L, or y and M. We make the interesting observation that if one propagates both left skew rays and right skew rays, the projected size of the propagating elliptical beam at any view angle φ is equal to the RMS average of the projected sizes of the left skew and right skew ray bundles (see Figure 3).

$$A(\varphi) = \sqrt{\frac{[A_R(\varphi)]^2 + [A_L(\varphi)]^2}{2}} \qquad (3)$$

Equation (3) can be proven to hold through any simply astigmatic optical system [7], and its validity for the case of general astigmatism was confirmed by comparing the calculated propagation of a generally astigmatic beam with previously published results [7].

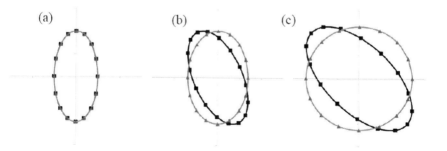

Figure 2. Evolution of a skew ray bundle corresponding to a 10 μm × 20 μm elliptical beam with λ = 1 μm (squares), compared with the actual beam profile (triangles), (a) at beam waist, (b) 300 μm from waist, and (c) 700 μm from waist.

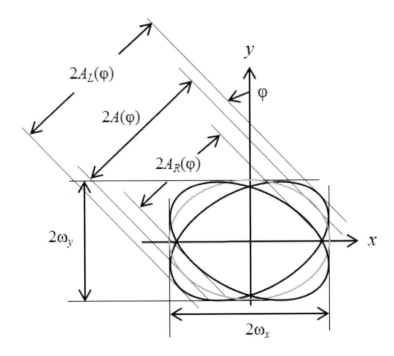

Figure 3. The projected size of the propagating elliptical beam $A(\varphi)$ is equal to the RMS average of the projected sizes of the left skew ray bundle $A_L(\varphi)$ and right skew ray bundle $A_R(\varphi)$.

Four Ray Representation

To calculate the Gaussian beam parameters through the optical system, it is only necessary to propagate two right skew rays to define the right skew ray bundle, and two left skew rays to define the left skew ray bundle. The four rays to be propagated are defined by

$$x_1 = p\omega_{0x}\cos(\alpha) \qquad x_3 = p\omega_{0x}\cos(\beta)$$
$$y_1 = p\omega_{0y}\sin(\alpha) \qquad y_3 = -p\omega_{0y}\sin(\beta)$$
$$L_1 = -p\theta_{0x}\sin(\alpha) \qquad L_3 = -p\theta_{0x}\sin(\beta)$$
$$M_1 = p\theta_{0y}\cos(\alpha) \qquad M_3 = -p\theta_{0y}\cos(\beta)$$
$$x_2 = p\omega_{0x}\cos(\alpha + \pi/2) \qquad x_4 = p\omega_{0x}\cos(\beta + \pi/2)$$

Representation of Gaussian Beams Using Rays

$$y_2 = p\omega_{0y}\sin(\alpha + \pi/2) \qquad y_4 = -p\omega_{0y}\sin(\beta + \pi/2)$$
$$L_2 = -p\theta_{0x}\sin(\alpha + \pi/2) \qquad L_4 = -p\theta_{0x}\sin(\beta + \pi/2)$$
$$M_2 = p\theta_{0y}\cos(\alpha + \pi/2) \qquad M_4 = -p\theta_{0y}\cos(\beta + \pi/2). \tag{4}$$

Ray 1 and ray 2 are right skew rays, and ray 3 and ray 4 are left skew rays. The pupil scaling factor p has been introduced to allow selection of rays closer to or further from the beam center, and the parameter β appears because an independent selection of rays can be used to represent the right and left skew ray bundles. The rays are propagated through the optical system in the normal way, and at a later point in the system we obtain new values of x, y, L, and M for each ray. For simplicity I use the same symbols for the new values. In addition, the central ray (ray 0) is propagated, launched on-axis from the center of the input beam, and all subsequent ray positions and angles are taken to be relative to the central ray. The subtraction can be done by the following process:

1. Subtract (x_0, y_0, z_0) from each ray coordinate (including ray 0).

$$x_{ib} = x_i - x_0$$
$$y_{ib} = y_i - x_0$$
$$z_{ib} = z_i - x_0 \tag{5}$$

2. Rotate all ray direction cosines and coordinates (including ray 0) about the y axis by an angle $\theta_y = \tan^{-1}(L_0/N_0)$.

$$x_{ic} = x_{ib}\cos\theta_y - z_{ib}\sin\theta_y$$
$$z_{ic} = z_{ib}\cos\theta_y + x_{ib}\sin\theta_y$$
$$L_{ic} = L_i\cos\theta_y - N_i\sin\theta_y$$
$$N_{ic} = N_i\cos\theta_y + L_i\sin\theta_y \tag{6}$$

3. Rotate all ray direction cosines and coordinates (including ray 0) about the x axis by an angle $\theta_x = \tan^{-1}(M_0/N_{0c})$.

$$y_{id} = y_{ib} \cos \theta_x - z_{ic} \sin \theta_x$$
$$z_{id} = z_{ic} \cos \theta_x + y_{ib} \sin \theta_x$$
$$M_{id} = M_i \cos \theta_x - N_{ic} \sin \theta_x$$
$$N_{id} = N_{ic} \cos \theta_x + M_i \sin \theta_x \tag{7}$$

4. Find the intercept of all rays with the plane $z = 0$.

$$x_{ie} = x_{ic} - \frac{L_{ic}}{N_{id}} z_{id}$$
$$y_{ie} = y_{id} - \frac{M_{id}}{N_{id}} z_{id} \tag{8}$$

The entire ray set should now have been transformed such that the central ray has position $(0,0,0)$ and direction cosines $(0,0,1)$, and all other ray positions and direction cosines will be relative to the central ray, with coordinates $(x_{ie}, y_{ie}, 0)$ in a plane perpendicular to the beam and direction cosines (L_{ic}, M_{id}, N_{id}). For simplicity, the step of calculating the ray positions and angles relative to the central ray is omitted from all following discussions. In Section 5, there are occasions where absolute ray coordinates are used (without the central ray subtracted) in the calculation of fiber coupling efficiency; those occasions are specifically noted in the text.

The projected size of the right skew ray bundle at a view angle φ is given by

$$A_R(\varphi) = \frac{1}{p} \sqrt{(x_1 \cos \varphi + y_1 \sin \varphi)^2 + (x_2 \cos \varphi + y_2 \sin \varphi)^2} \tag{9}$$

and the projected size of the left skew ray bundle is given by

$$A_L(\varphi) = \frac{1}{p} \sqrt{(x_3 \cos \varphi + y_3 \sin \varphi)^2 + (x_4 \cos \varphi + y_4 \sin \varphi)^2} \tag{10}$$

and therefore using Equation (3) the projected size of the propagating Gaussian beam is given by

Representation of Gaussian Beams Using Rays   11

$$A(\varphi) = \frac{1}{p}\sqrt{\frac{(x_1\cos\varphi+y_1\sin\varphi)^2+(x_2\cos\varphi+y_2\sin\varphi)^2+(x_3\cos\varphi+y_3\sin\varphi)^2+(x_4\cos\varphi+y_4\sin\varphi)^2}{2}} \ . \quad (11)$$

If all one requires is the projected size of the Gaussian beam at a particular view angle, for example for the purpose of drawing the Gaussian beam on a layout diagram, Equation (11) is a very computationally efficient way to get that information.

The choice of rays to be launched is not limited to the definition of Equation (4). The x and y components are independent, and therefore a "phase difference" γ between x and y components can be introduced without affecting the calculated value of Equation (11). This creates left and right skew ray bundles not coincident with the intensity ellipse, as in Figure 3, even at the beam waist. But even this is not the complete set of possible launched rays. If the starting beam is circular, it can be easily seen that ray bundles rotated about the z axis should also be valid representations of the beam. Similarly for elliptical starting beams, other ray bundles may be used although they are not simple rotations. The ray definitions including such rotations are given by Equation (12). Equation (12) defines ray 1, and substitutions are made for α and γ according to Table 1 to produce rays 2, 3, and 4.

$$x = p\omega_{0x}[\cos\alpha\cos\delta + \sin(\alpha+\gamma)\sin\delta]$$
$$y = p\omega_{0y}[\sin(\alpha+\gamma)\cos\delta - \cos\alpha\sin\delta]$$
$$L = -p\theta_{0x}[\sin\alpha\cos\delta - \cos(\alpha+\gamma)\sin\delta]$$
$$M = p\theta_{0y}[\cos(\alpha+\gamma)\cos\delta + \sin\alpha\sin\delta] \quad (12)$$

Table 1. Substitutions to be made in Equation (12) to produce the four rays used to compute Gaussian beam parameters

	α	γ
Ray 1	α	γ
Ray 2	$\alpha + \pi/2$	γ
Ray 3	β	$\gamma + \pi$
Ray 4	$\beta + \pi/2$	$\gamma + \pi$

Equation (12) and Table 1 encompass all possible sets of rays which can be used with the analysis methods of this chapter. Any values of p, α, β, γ, and δ may be selected and the resulting four rays will represent the propagating beam and produce the same calculated results of the beam parameters (see Section 3), assuming no aberrations.

New parameters γ and δ have been introduced which change the shapes of the ray bundles, as shown in Figure 4. Figure 4 shows the locations of rays produced by Equation (12) for 16 values of α and β, for selected values of γ and δ, and $p = 1$. In each case, the same elliptical beam is being represented (the solid oval), and in each case Equation (3) holds for all view angles.

In general, if a ray bundle (that is, rays with all possible values of α and β) corresponding to particular values of γ and δ is launched through an optical system, and the beam is subsequently re-focused at the output, the shape of the resulting ray bundle at intermediate points or at the output will correspond to different values of γ and δ. Supposing one is using such a ray bundle to optimize the optical system, one might launch skew rays with $\gamma = 0$ or π and a range of values of α and p which sample the entire area of the corresponding Gaussian beam, but end up with rays distributed only along two diagonal lines (like the ray bundles corresponding to $\gamma = \pi/2$ or $3\pi/2$). Rays concentrated on two diagonal lines may not be sufficient for optimizing certain system aberrations such as astigmatism. To avoid this potential problem, one can launch multiple ray bundles with different values of γ and δ to ensure that the distribution of rays is always sufficient to sample the entire region of the optical aperture which is covered by the propagating beam.

The parameters γ and δ are somewhat analogous to polarization states in terms of their effect on the shapes of the ray bundles. When $\gamma = 0$, the right skew and left skew ray bundles are analogous to right circular and left circular polarization states. When $\gamma = \pi/2$ or $3\pi/2$, the ray bundles correspond to linear states of polarization; $\delta = 0$ corresponds to $\pm 45°$ linearly polarized states and $\delta = \pi/4$ corresponds to $0°$ and $90°$ polarized states. A suitable choice of ray bundles to ensure complete sampling of the

Representation of Gaussian Beams Using Rays 13

optical aperture under all conditions would be (1) γ = 0 and π, δ = 0, (2) γ = π/2 and 3π/2, δ = 0, and (3) γ = π/2 and 3π/2, δ = π/4.

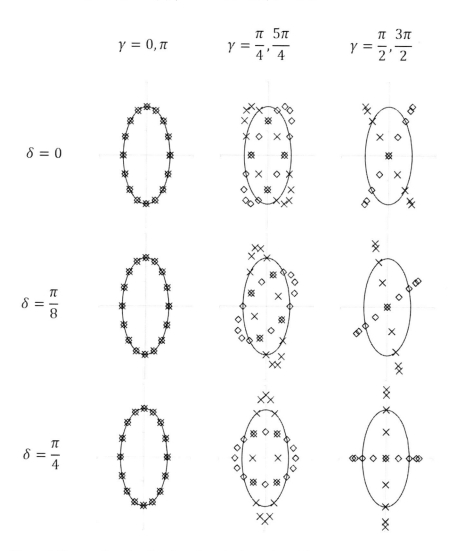

Figure 4. Shapes of ray bundles for selected values of γ and δ. In each case the same elliptical beam shape (solid oval) is being represented, and ray positions for 16 values of α or β are shown. ◊ indicates the ray positions for the smaller value of γ and × indicates the larger value of γ.

14 Paul D. Colbourne

Ray Density Proportional to Amplitude or Intensity

The analogy between ray bundles and polarization states raises the question: what if a ray bundle is constructed which includes rays with values of γ and δ corresponding to all points on a polarization sphere (Poincaré sphere)? That is, we might randomly select a point on a sphere, assign values of γ and δ corresponding to that point, select a random value of α, and generate a ray using Equation (12), repeating to create a large number of rays. If U_i are random variables uniformly distributed between 0 and 1, values of α, γ, and δ could be assigned using

$$\alpha = 2\pi U_1$$
$$\gamma = \cos^{-1}(2U_2 - 1)$$
$$\delta = \pi U_3. \tag{13}$$

Defining rays using Equation (13) and $p = 1$ results in rays that are uniformly distributed in space up to $\sqrt{2}\omega_0$ and uniformly distributed in angle up to $\sqrt{2}\theta_0$, as shown in Figure 5. I will not give in detail here the proof that the rays are uniformly distributed, but the proof involves calculating $v = x^2/\omega_{0x}^2 + y^2/\omega_{0y}^2$ as a function of U_1 and U_2, and showing by integration that in the space defined by U_1 and U_2, where the rays are uniformly distributed by definition, the portion of the rays with v less than a threshold v_0 is proportional to v_0; therefore, the number of rays with $v < v_0$ in the physical space (x,y) is proportional to the enclosed area and so the distribution is uniform.

Because each ray is a member of the ray set which represents the propagating Gaussian beam, if we propagate the complete ray bundle, the rays maintain a uniform density in space up to $\sqrt{2}\omega$ at all points along the beam propagation. If we also select random values of p with an appropriate distribution, we can generate a set of rays which have a ray density at all points corresponding to the field amplitude or the intensity of the propagating Gaussian beam. This can be useful for analyzing stray light, using features built into optical ray tracing programs which scatter rays based on surface roughness, reflectivity, fluorescence, etc. The appropriate distribution of p for creating rays distributed according to the field amplitude is

$$P(p) = 8p^3 e^{-2p^2}. \tag{14}$$

Equation (14) was derived by imagining a stack of cylinders of radius $r = \sqrt{2}p$ and height dh which fill the space under the normalized Gaussian field profile $f(r) = (1/\pi)e^{-r^2}$. The weight we want to put on each value of p is equal to the volume of the corresponding cylinder.

$$dV = \pi r^2 \cdot dh = -\pi r^2 \frac{df(r)}{dr} dr = 2r^3 e^{-r^2} dr = 8p^3 e^{-2p^2} dp \tag{15}$$

One method of creating random values of p with the distribution of Equation (14) is to do a root of sum of squares average of four normally distributed random variables X_1, X_2, X_3, and X_4, each with mean equal to 0 and standard deviation equal to 1.

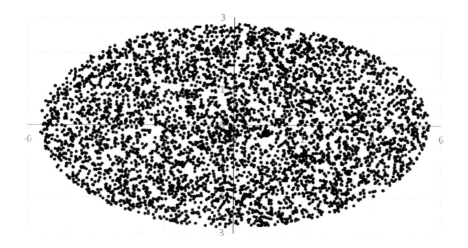

Figure 5. Plot of (x,y) values for 5000 random rays, using Equation (13) to generate values of α, γ, and δ to insert into Equation (12), and $p = 1$, with $\omega_{0x} = 4$ μm and $\omega_{0y} = 2$ μm.

$$p = \frac{1}{2}\sqrt{X_1^2 + X_2^2 + X_3^2 + X_4^2} \qquad (16)$$

The Box-Muller transform is a method of generating normally distributed random numbers [10]. If U and V are random variables uniformly distributed between 0 and 1, and we assign

$$X_1 = \sqrt{-2\ln U}\cos 2\pi V$$
$$X_2 = \sqrt{-2\ln U}\sin 2\pi V \qquad (17)$$

then X_1 and X_2 are normally distributed random variables. Since we are going to square and add X_1 and X_2, we have

$$X_1^2 + X_2^2 = -2\ln U. \qquad (18)$$

Similarly we are going to square and add X_3 and X_4, so inserting Equation (18) twice into Equation (16) we have

$$p = \frac{1}{2}\sqrt{-2\ln U_4 - 2\ln U_5}\ . \qquad (19)$$

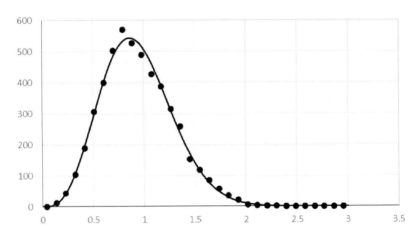

Figure 6. A histogram of 5000 random values of p (markers) generated using Equation (19), along with the desired distribution of p given by Equation (14) (solid line).

Representation of Gaussian Beams Using Rays

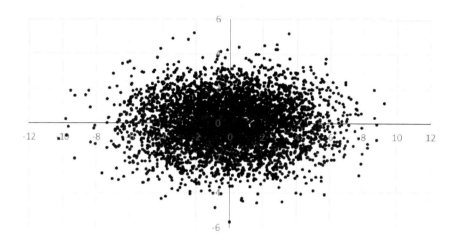

Figure 7. Plot of (*x*,*y*) values for 5000 random rays generated using Equation (13) and *p* given by Equation (19), with ω_{0x} = 4 μm and ω_{0y} = 2 μm.

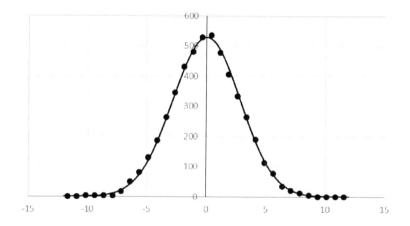

Figure 8. A histogram (markers) of the values of *x* of the ray distribution in Figure 7, along with a calculation of the corresponding Gaussian field distribution (solid line).

Figure 6 shows a histogram of 5000 random values of *p* generated using Equation (19), and the distribution matches closely with Equation (14). Using *p* defined by Equation (19) and α, γ, and δ defined by Equation (13) gives us a way of generating a random set of rays which will have a ray density proportional to the Gaussian field at all points in the optical system, as shown in Figures 7-9. Figure 9 shows that after propagating the rays away from the waist, the distribution of rays continues to match the Gaussian field distribution of the propagating Gaussian beam. To create rays distributed according to the intensity, the values of p would be chosen smaller than Equation (19) by a factor of $\sqrt{2}$. Such a ray set can be used in a Monte Carlo analysis to calculate stray light in an optical system which uses Gaussian beams.

3. CALCULATION OF BEAM PROPERTIES

A generally astigmatic Gaussian beam has an elliptical intensity distribution, and elliptical or hyperbolic equal-phase contours which are not necessarily aligned with the intensity ellipse [11]. We have shown that the projected size of the Gaussian beam at any view angle can be calculated from four propagated skew rays. Now the task is to use this information to determine the orientation and principal axes of the intensity ellipse, and the orientation and curvatures along the principal axes of the wavefront surface.

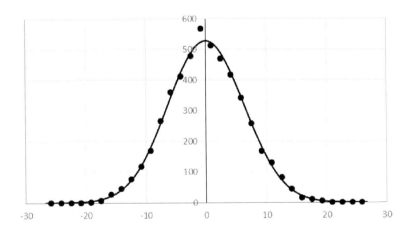

Figure 9. A histogram (markers) of the values of x of the ray distribution in Figure 7, after the rays were propagated to a distance $z = 100$ μm, along with a calculation of the expected Gaussian field distribution at $z = 100$ μm (solid line). A wavelength of $\lambda = 1$ μm was used in the ray generation and the Gaussian field calculation.

Intensity Ellipse

The orientation of the intensity ellipse can be determined by solving

$$\frac{dA}{d\varphi} = 0. \tag{20}$$

Replacing Equation (11) into Equation (20) we get

$$\frac{1}{p^2 A}[(x_1 \cos\varphi + y_1 \sin\varphi)(-x_1 \sin\varphi + y_1 \cos\varphi)$$
$$+(x_2 \cos\varphi + y_2 \sin\varphi)(-x_2 \sin\varphi + y_2 \cos\varphi)$$
$$+(x_3 \cos\varphi + y_3 \sin\varphi)(-x_3 \sin\varphi + y_3 \cos\varphi)$$
$$+(x_4 \cos\varphi + y_4 \sin\varphi)(-x_4 \sin\varphi + y_4 \cos\varphi)] = 0. \tag{21}$$

Equation (21) rearranges to

$$[x_1 y_1 + x_2 y_2 + x_3 y_3 + x_4 y_4](\cos^2\varphi - \sin^2\varphi)$$
$$+[y_1^2 + y_2^2 + y_3^2 + y_4^2 - x_1^2 - x_2^2 - x_3^2 - x_4^2](\cos\varphi\sin\varphi) = 0. \tag{22}$$

Making use of the trigonometric identities

$$\cos^2\varphi - \sin^2\varphi = \cos 2\varphi \tag{23}$$

$$\cos\varphi\sin\varphi = \frac{\sin 2\varphi}{2} \tag{24}$$

we can then solve for the orientation of the intensity ellipse:

$$\varphi' = \frac{1}{2}\tan^{-1}\left(\frac{2(x_1 y_1 + x_2 y_2 + x_3 y_3 + x_4 y_4)}{y_1^2 + y_2^2 + y_3^2 + y_4^2 - x_1^2 - x_2^2 - x_3^2 - x_4^2}\right). \tag{25}$$

If the denominator of Equation (25) is equal to zero, we set $\varphi' = \pi/4$. We can define new coordinate axes (x', y') aligned with the intensity ellipse, to simplify the remaining calculations. All ray positions and angles can be mapped to the new coordinates using

$$x'_1 = x_1\cos(\varphi') + y_1\sin(\varphi')$$
$$y'_1 = -x_1\sin(\varphi') + y_1\cos(\varphi')$$
$$L'_1 = L_1\cos(\varphi') + M_1\sin(\varphi')$$
$$M'_1 = -L_1\sin(\varphi') + M_1\cos(\varphi'). \tag{26}$$

The principal axes $\omega_{x'}$ and $\omega_{y'}$ of the intensity ellipse are then given by

$$\omega_{x'} = \frac{1}{p}\sqrt{\frac{{x'_1}^2 + {x'_2}^2 + {x'_3}^2 + {x'_4}^2}{2}}$$
$$\omega_{y'} = \frac{1}{p}\sqrt{\frac{{y'_1}^2 + {y'_2}^2 + {y'_3}^2 + {y'_4}^2}{2}}. \tag{27}$$

Assuming that the ray coordinates have been determined in a plane perpendicular to the beam, Equation (27) gives the beam size in that perpendicular plane. In some cases, we may want instead the size of the beam projected onto an optical surface, for example to determine whether the beam falls within the clear aperture of a tilted fold mirror. In such a case, the ray coordinates in the surface coordinates may be used in Equations (25-27).

Wavefront Curvature

We can determine phase information by calculating dA/dz. The value of dA/dz corresponds to the slope of the wavefront surface in a direction perpendicular to the intensity ellipse, at the tangent point of the intensity ellipse, since the propagation direction is perpendicular to the wavefront. If we calculate $(dA/dz)/A$, this gives the curvature of the wavefront along the direction φ (see Figure 10 (a)), provided that a line perpendicular to the intensity ellipse passes through the origin. This condition will be satisfied along the major and minor axes of the intensity ellipse, as shown in Figure 10 (b). Therefore, we have that the wavefront curvatures along x' and y' are

$$C_{x'} = \frac{\left(\frac{d\omega_{x'}}{dz}\right)}{\omega_{x'}} = \frac{x'_1 l'_1 + x'_2 l'_2 + x'_3 l'_3 + x'_4 l'_4}{x'^2_1 + x'^2_2 + x'^2_3 + x'^2_4}$$

$$C_{y'} = \frac{\left(\frac{d\omega_{y'}}{dz}\right)}{\omega_{y'}} = \frac{y'_1 m'_1 + y'_2 m'_2 + y'_3 m'_3 + y'_4 m'_4}{y'^2_1 + y'^2_2 + y'^2_3 + y'^2_4}. \tag{28}$$

Note that in Equation (28) the ray slopes $l = dx/dz = L/N$ and $m = dy/dz = M/N$ are used, not the direction cosines L and M, because Equation (28) derives from dA/dz. We can also calculate dA/dz at an angle $\pi/4$ from the x' axis.

$$\frac{dA'(\pi/4)}{dz} = \frac{(x'_1 + y'_1)(l'_1 + m'_1) + (x'_2 + y'_2)(l'_2 + m'_2) + (x'_3 + y'_3)(l'_3 + m'_3) + (x'_4 + y'_4)(l'_4 + m'_4)}{2p\sqrt{(x'_1 + y'_1)^2 + (x'_2 + y'_2)^2 + (x'_3 + y'_3)^2 + (x'_4 + y'_4)^2}} \tag{29}$$

To complete the description of the propagating generally astigmatic Gaussian beam, we need to determine the orientation of the equal-phase contours and the curvatures along the principal axes of the wavefront surface. We can extract this information from the calculated values of $C_{x'}$, $C_{y'}$, and $dA'(\pi/4)/dz$ as follows. We know that in coordinates (x'',y'') oriented along the principal axes of the wavefront surface, the sag of the wavefront surface can be defined as (for small values of x'' and y''):

$$S = \frac{C_{x''} x''^2 + C_{y''} y''^2}{2}. \tag{30}$$

In the coordinate system (x',y'), rotated φ'' from the coordinates (x'',y''), the wavefront surface becomes

$$S = \frac{C_{x'} x'^2 + C_{y'} y'^2}{2} + b x' y'. \tag{31}$$

The tangent point to the intensity ellipse at an angle $\pi/4$ is given by

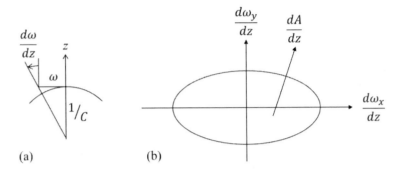

Figure 10. (a) Illustration of how *dA/dz* provides wavefront curvature information. The wavefront is perpendicular to the propagation direction, and the propagation direction follows the boundary of the intensity ellipse. (b) The calculation of *dA/dz* directly relates to the wavefront curvature *C*, as in (a), only when a line perpendicular to the intensity ellipse passes through the center of the ellipse; this occurs only along the major and minor axes ($d\omega_x/dz$ and $d\omega_y/dz$).

$$u = \tan^{-1}\left(\frac{\omega_{y'}}{\omega_{x'}}\tan\frac{\pi}{4}\right)$$
$$x'_t = \omega_{x'}\cos(u)$$
$$y'_t = \omega_{y'}\sin(u). \tag{32}$$

We can calculate *dA/dz* (numerically equal to the slope of the wavefront surface in a direction perpendicular to the intensity ellipse) using Equation (31):

$$\frac{dA'(\pi/4)}{dz} = \left[\frac{dS}{dx'}\cos\frac{\pi}{4} + \frac{dS}{dy'}\sin\frac{\pi}{4}\right]_{(x'_t,y'_t)} = \frac{x'_t C_{x'} + b(x'_t + y'_t) + y'_t C_{y'}}{\sqrt{2}} \tag{33}$$

and rearranging, we obtain the value of *b* as

$$b = \frac{\sqrt{2}\frac{dA'(\pi/4)}{dz} - x'_t C_{x'} - y'_t C_{y'}}{x'_t + y'_t} \tag{34}$$

where the value of *dA'(π/4)/dz* can be determined from Equation (29), and $C_{x'}$, $C_{y'}$, x'_t, and y'_t are determined from Equations (28) and (32). Now all constants of Equation (31) have been determined, so we have a complete expression for the wavefront surface, and we have only to determine the coordinate axis orientation for which the wavefront expression reduces to Equation (30). To determine the orientation, we calculate the wavefront sag around a unit circle. Three points are sufficient to determine the orientation of the wavefront surface. We choose points at angles 0, $\pi/4$, and $\pi/2$ around the unit circle.

$$S_0 = \frac{C_{x'}}{2}$$
$$S_{\pi/2} = \frac{C_{y'}}{2}$$
$$S_{\pi/4} = \frac{C_{x'} + C_{y'}}{4} + \frac{b}{2} \tag{35}$$

We calculate the quantities

$$R = \frac{S_0 + S_{\pi/2}}{2}$$
$$T = \sqrt{(S_0 - R)^2 + \left(S_{\pi/4} - R\right)^2}. \tag{36}$$

The orientation of the wavefront surface relative to the intensity ellipse is then given by

$$\varphi'' = \frac{\text{atan2}\left(\frac{S_0 - R}{T}, \frac{S_{\pi/4} - R}{T}\right)}{2} \tag{37}$$

where atan2(x,y) is an arctangent function commonly found in math libraries which can return angles from 0 to 2π. The curvatures along the major and minor axes are

$$C_{x''} = 2(R + T)$$
$$C_{y''} = 2(R - T). \tag{38}$$

24 *Paul D. Colbourne*

We now have the complete description of the propagating generally astigmatic Gaussian beam, extracted from the propagation of four skew rays. This calculation approach is not necessarily more computationally efficient than the matrix methods described by Arnaud [5] or Lü et al. [6], however it has been derived without reference to physically unrealistic concepts such as imaginary rays or imaginary rotation angles, and therefore has some value in furthering understanding of the propagation of generally astigmatic Gaussian beams.

Minimum Projected Beam Size

A generally astigmatic Gaussian beam has no point which can be identified as the "waist" in the same sense as a simply astigmatic Gaussian beam, however one useful parameter which can be defined is the location of the minimum projected beam size for a given view angle. For example, if we view in a direction parallel to the y axis, there exists one value of z for which the beam width in the x direction is a minimum. We find this value of z by solving

$$\frac{dA(0)}{dz} = 0 \tag{39}$$

which gives us

$$z_x = \frac{x_1 l_1 + x_2 l_2 + x_3 l_3 + x_4 l_4}{l_1^2 + l_2^2 + l_3^2 + l_4^2}. \tag{40}$$

Note that in Equation (40) the ray slope $l = dx/dz = L/N$ is used, not the direction cosine L, because Equation (40) derives from dA/dz. Once the distance z_x is known, it is straightforward to extrapolate the ray positions to that point in z using Equation (8) and then calculate the beam size at that position using Equation (11). To calculate the location of the minimum projected beam size for arbitrary view angles φ, replace x_i in Equation (40) with $(x_i \cos\varphi + y_i \sin\varphi)$, and replace l_i with $(l_i \cos\varphi + m_i \sin\varphi)$.

Divergence

Another useful parameter is the beam divergence D, which can be calculated in a manner analogous to Equation (11).

$$D(\varphi) = \frac{1}{p}\sqrt{\frac{(L_1\cos\varphi+M_1\sin\varphi)^2+(L_2\cos\varphi+M_2\sin\varphi)^2+(L_3\cos\varphi+M_3\sin\varphi)^2+(L_4\cos\varphi+M_4\sin\varphi)^2}{2}}. \quad (41)$$

We use direction cosines L and M in Equation (41) so that we get the same value of divergence regardless of the value of p chosen. The divergence will have minimum and maximum values as φ is varied, in the same way as the projected beam size A, and the orientation of the maximum or minimum value of D can be found in the same way as the orientation of the intensity ellipse.

$$\varphi_D = \frac{1}{2}\tan^{-1}\left(\frac{2(L_1M_1+L_2M_2+L_3M_3+L_4M_4)}{M_1^2+M_2^2+M_3^2+M_4^2-L_1^2-L_2^2-L_3^2-L_4^2}\right) \quad (42)$$

Once φ_D is known, it and $\varphi_D + \pi/2$ can be substituted for φ in Equation (41) to calculate the minimum and maximum divergence.

The orientation of the maximum divergence may be useful as a reference for calculating other properties of generally astigmatic Gaussian beams, for example to return values which will degenerate into the usual waist positions and sizes when the beam is simply astigmatic. However, φ_D does not help in identifying a "waist" of a generally astigmatic beam, because the minimum projected beam size at a view angle φ_D is not in general the smallest or largest projected beam size over all view angles.

4. OPTIMIZATION USING SKEW RAYS

A ray optimization of a system which couples light from one fiber to another, such as the system shown in Figure 1, would normally be done using rays diverging from a point at the input (object) surface and converging to a point at the output (image) surface. The optimization process would involve altering the system parameters (lens radii of curvature, distances between elements, etc.) until the spot size created by the ray positions at the output surface is minimized. A ray pattern with multiple rings and arms would be propagated, where each ring consists of rays with the same angle from the central ray, and each arm consists of rays with the same azimuthal angle. Using multiple rings and arms ensures that the optical aperture is sufficiently sampled to enable optimization of the various aberrations that might be present. However, using rays diverging from a point does not give the correct Gaussian beam focus. One simple example is a collimating lens, which will produce parallel rays if all incoming rays originate at the focal point of the lens. The parallel rays will maintain the same cross-section no matter how far they are propagated. A collimated Gaussian beam, on the other hand, will eventually diverge, no matter the size of the collimated beam. Re-focusing the parallel rays therefore produces incorrect results if the propagation distance is sufficient for there to be significant divergence of a Gaussian beam. What is needed is an optimization method which can minimize aberrations, but also give the correct Gaussian beam focus.

Generally Astigmatic Systems

We can accomplish our goal by launching skew rays as in Equation (12) instead of rays diverging from a point, because the rays defined by Equation (12) will propagate in accordance with the propagation of a generally astigmatic Gaussian beam. One difficulty is that when the beam is in focus, the rays at the output do not converge to a point, so the spot size created by the output rays is not a suitable parameter for optimization. We note that the ray positions for ray 1 (using α in Equation (12)) are proportional to the ray direction cosines for ray 2 (using $\alpha + \pi/2$ in Equation (12)).

Representation of Gaussian Beams Using Rays

$$L_2 = -p\theta_{0x}[\sin(\alpha + \pi/2)\cos\delta - \cos(\alpha + \pi/2 + \gamma)\sin\delta]$$
$$= -p\theta_{0x}[\cos\alpha\cos\delta + \sin(\alpha + \gamma)\sin\delta] = -(\theta_{0x}/\omega_{0x})x_1$$
$$M_2 = p\theta_{0y}[\cos(\alpha + \pi/2 + \gamma)\cos\delta + \sin(\alpha + \pi/2)\sin\delta]$$
$$= p\theta_{0y}[\sin(\alpha + \gamma)\cos\delta + \cos\alpha\sin\delta] = -(\theta_{0y}/\omega_{0y})y_1 \tag{43}$$

This suggests that we can use the ray direction cosines of ray 2 to calculate an offset to apply to ray 1, so that the resulting position of ray 1 lands at the beam center, if the beam is in focus. Since $\omega_0/\theta_0 = z_R$, the Rayleigh range, we can calculate the offset ray positions using

$$x_{1b} = x_1 + L_2 z_{RxT}$$
$$y_{1b} = y_1 + M_2 z_{RyT}. \tag{44}$$

Here z_{RxT} and z_{RyT} are the Rayleigh ranges of the target output beam waist in x and y. In this way we can get the offset positions of all rays to land at the same point when the beam is in focus, and the spot size created by the offset rays provides a suitable parameter for optimization since it tends to zero at perfect focus. To work properly with generally astigmatic optical systems (that is, no limitations on the orientations of optical elements), rays for both γ and $\gamma + \pi$ need to be propagated and included in the spot size calculation. Note that the input beam and the output beam must be elliptical with both the x and y waists located at the input or output surface. If the desired input and output beams do not fit this condition, lenses or coordinate breaks may be placed at the input or output to convert the desired input or output beam into a beam meeting the required condition.

This optimization method was implemented in Zemax (optical design software available from Zemax, LLC) using an optimization merit function macro which propagates three rings (with different values of p) and eight arms of rays (the arms having different values of α equally spaced from 0 to 2π, and the number of arms divisible by 4 so that for each ray, a ray at $\alpha + \pi/2$ is available) as in Equation (12) with $\gamma = \delta = 0$ plus a second set of three rings and eight arms with $\gamma = \pi$, offsets the position of each ray at the output according to Equation (44), and returns an RMS spot size calculated

28 Paul D. Colbourne

using the offset ray positions. At the time of writing, a similar macro called ZPL31.ZPL which operates based on ray slopes instead of direction cosines (there is negligible difference for low divergence beams with NA < 0.25) was available on the recorded webinars page on the www.zemax.com web site, under the webinar "Using Skew Rays to Model Gaussian Beams," as part of the ".zar" archive files posted there.

If the optimization is successful and the RMS spot radius is forced to zero, then

$$x_1 = -L_2 \omega_{0xT}/\theta_{0xT}$$
$$y_1 = -M_2 \omega_{0yT}/\theta_{0yT} \tag{45}$$

and if we assume that the ray at $\alpha + \pi$ has position and angle equal in magnitude but opposite in sign to those of ray 1 (which follows due to symmetry if there are no aberrations),

$$x_2 = L_1 \omega_{0xT}/\theta_{0xT}$$
$$y_2 = M_1 \omega_{0yT}/\theta_{0yT}. \tag{46}$$

Translating Equations (42) and (43) of Herloski et al. [3] into the notation of this paper, for the x component only, we have that for the case of simple astigmatism

$$\omega_x = \frac{1}{p}\sqrt{{x_1}^2 + {x_2}^2} \tag{47}$$

$$z = \frac{x_1 l_1 + x_2 l_2}{{l_1}^2 + {l_2}^2}. \tag{48}$$

In Equation (47) the factor $1/p$ has been inserted since in Equation (12) the definition of launched rays allows the pupil scaling factor p. Herloski et al. also state in words in the paragraphs following their Equations (42) and (43) that

Representation of Gaussian Beams Using Rays

$$\theta_{0x} = \frac{1}{p}\sqrt{l_1^2 + l_2^2} \tag{49}$$

where again the factor $1/p$ has been inserted. Using Equations (45) and (46) to replace for x_1 and x_2 in Equation (48), and making the approximation that $l=L$ and $m=M$, we get the result that the beam waist is located at the output surface ($z = 0$), and therefore Equation (47) will give us the waist size ω_{0x} of the output beam. Using Equations (47) and (49) to calculate the Rayleigh range squared $z_{Rx}^2 = \omega_{0x}^2/\theta_{0x}^2$ and substituting for x_1 and x_2 using Equations (45) and (46), we get the result that $\omega_{0x}^2/\theta_{0x}^2 = \omega_{0xT}^2/\theta_{0xT}^2$ and therefore the actual beam radius ω_{0x} is equal to the desired waist size ω_{0xT}. A similar analysis can be done for the y component. We have thus proven that at least for the case of simple astigmatism where Equations (47) to (49) apply, the desired result of a beam with waists ω_{0xT} and ω_{0yT} located at the output surface has been achieved. Given that propagation of rays with γ and $\gamma + \pi$ is sufficient to represent a generally astigmatic Gaussian beam, obtaining an RMS spot size near zero at the output for both γ and $\gamma + \pi$ ensures that even with intervening general astigmatism the correct focus is achieved. Testing with several examples shows this to be the case [7].

A key feature of this optimization method is the computational simplicity. The only calculations required beyond that of a conventional ray optimization are the ray definitions of Equation (12) and the calculation of Equation (44). There is no need to compute the waist size and position of the propagating beam and compare them with desired values.

Circular Beams

If the input beam is circular ($\omega_{0x} = \omega_{0y} = \omega_0$) and remains circular throughout the optical system, there is a particularly simple method of performing optimization using skew rays. This was implemented in Zemax by creating a user-defined surface which offsets rays according to their

30 Paul D. Colbourne

angle and the beam waist radius ω_0 which is entered as a parameter of the user-defined surface.

$$x = x + M\omega_0/|\theta_0|$$
$$y = y - L\omega_0/|\theta_0| \tag{50}$$

(A similar user-defined surface called "us_gskew.dll" was available on the recorded webinars page at www.zemax.com at the time of writing, but us_gskew.dll operates on ray slopes, not direction cosines. The difference is negligible for low divergence systems with NA < 0.25.) It can be verified that for rays diverging from a point, the user-defined surface placed at the input converts each launched ray to a skew ray meeting the conditions of Equation (12) with $\gamma = \delta = 0$, and appropriate values of p and α or β. The absolute value of θ_0 is used in Equation (50) so that by specifying a negative value of ω_0, the sign of the ray offset can be reversed. A second user-defined surface can be placed at the output surface with the opposite sign of ω_0 (but not necessarily the same magnitude as the input value if the desired output beam radius is different from the input beam radius). This reverses the skew ray offsets that were created at the input, resulting in output rays that converge to a point if the beam is focused at the output surface. Then, conventional ray optimization algorithms can be used to optimize the system and it will converge such that the Gaussian beam is focused. However, this method fails if the system contains cylindrical lenses, even if the beam returns to a circular beam at the output. The reason is that in general the output skew ray bundle will correspond to a non-zero value of γ after passing through cylindrical lenses, so implementing Equation (50) at the output does not result in rays converging to a point. However, the convenience of the user-defined surface approach makes the use of Equation (50) attractive for the quite common case where the beam is circular. The method of Equation (44) cannot be implemented using a user-defined surface because the user-defined surface operates on only one ray at a time.

Large Aberrations

The equivalence between rays and Gaussian beams (and indeed Gaussian beam propagation theory itself) is valid only when there are no aberrations present, but presumably the final optimized state has negligible aberrations if an RMS spot radius near zero is achieved. If the RMS spot radius is much larger than the desired spot size, the ray offsets introduced in the optimization process are negligible and the optimization is essentially equivalent to a conventional ray optimization. It is only when aberrations become small, and the optical system approaches being diffraction limited, that the ray offsets become significant, and in this state the equivalence between rays and Gaussian beams is valid.

Sufficient Sampling

As previously mentioned in Section 2, the shapes of the ray bundles can change as the rays propagate through the optical system. If one launches rays corresponding to only one value of γ and one value of δ, one might end up with rays distributed only along two diagonal lines, as in the figures corresponding to $\gamma = \pi/2$ or $3\pi/2$ in Figure 4. Rays concentrated on two diagonal lines may not be sufficient for optimizing certain aberrations such as astigmatism. To avoid this potential problem, one can launch multiple ray bundles with different values of γ and δ to ensure that the distribution of rays is sufficient to sample the entire region of the optical aperture which is covered by the propagating beam. A suitable choice of ray bundles to ensure complete sampling of the optical aperture under all conditions would be (1) $\gamma = 0$ and π, $\delta = 0$, (2) $\gamma = \pi/2$ and $3\pi/2$, $\delta = 0$, and (3) $\gamma = \pi/2$ and $3\pi/2$, $\delta = \pi/4$ (top left, top right, and bottom right in Figure 4, respectively). These rays could be launched with only a single value of p, to avoid an excessive number of traced rays, since the rays of case (2) and case (3) will include rays at various distances from the optical axis. Of course, more rays can be launched if desired, for example using multiple values of p or other values of γ and δ, perhaps corresponding to the vertices

32   *Paul D. Colbourne*

of a polyhedron such as a dodecahedron on the polarization sphere. (The rays of cases (1), (2), and (3) above correspond to the six vertices of an octahedron on the polarization sphere.)

5. COUPLING EFFICIENCY CALCULATION

In Section 2, a method was described for generating a set of rays with density corresponding to the field of a propagating Gaussian beam. Can such a ray set be used to compute the output field, thus enabling calculation of the coupling efficiency for the output beam into a mode such as an optical fiber or a waveguide? In this section I describe a method similar to that described by Wagner and Tomlinson [12], but using the ray bundle representing a Gaussian beam instead of the conventional rays diverging from a point. With this method, the coupling efficiency can be calculated with the proper Gaussian beam focus, subject to the same limitations as the Wagner and Tomlinson approach, the main limitation being that the aberrations must be in the far-field (that is, much further from the output than the Rayleigh range of the output mode). We use the angular spectrum approach to compute the output field, whereby an infinite plane wave is associated with each ray, and the plane waves are summed at the output to give the output field distribution.

Ray Generation

The first step is to generate an appropriate set of rays. Using Equations (13) and (19) would produce a random set of rays with the appropriate ray density. However, random rays may not be the best choice, since calculation noise would cause different results each time the coupling efficiency is calculated, making it difficult to optimize an optical system based on the coupling efficiency results. A better option would be to select values of α uniformly spaced from 0 to 2π (the number of values of α should be a multiple of 4 because, as with the optimization method, we

Representation of Gaussian Beams Using Rays 33

will need rays with $\alpha + \pi/2$ for offset calculations), and values of γ and δ corresponding to equally spaced points on the polarization sphere. Suitable sets of points on the sphere would be the 12 vertices of an icosahedron, the 20 vertices of a dodecahedron, the combined 32 vertices of the icosahedron and dodecahedron (since the icosahedron vertices lie at the midpoints of the dodecahedron faces and vice versa) (these are the vertices of a "pentakis dodecahedron"), or the 32 pentakis dodecahedron vertices plus the 30 midpoints between icosahedron vertices (these 62 points are the vertices of a "disdyakis triacontahedron"). The icosahedron vertices are (x,y,z) = circular permutations of $(0, \pm 1, \pm \phi)$, where $\phi = (1 + \sqrt{5})/2$ is the "golden ratio." The dodecahedron vertices are $(x,y,z) = (\pm 1, \pm 1, \pm 1)$ and circular permutations of $(0, \pm \phi, \pm 1/\phi)$. Once the vertices are selected, the values of γ and δ are assigned using

$$\gamma = \text{atan2}\left(z, \sqrt{x^2 + y^2}\right)$$
$$\delta = \frac{\text{atan2}(x,y)}{2} \tag{51}$$

where $\text{atan2}(x,y)$ is an arctangent function which can return values between 0 and 2π.

Instead of using Equation (19) to select random values of p, we would like a deterministic way of producing a set of values of p satisfying Equation (14). If we want a list of N_p values of p, we can assign the first member of the list using

$$p_1 = \sqrt[4]{\frac{1}{4N_p}}. \tag{52}$$

Subsequent members of the list could be calculated using

$$p_i = p_{i-1} + \frac{1}{8N_p p^3 e^{-2p^2}} \tag{53}$$

where p would be the average of p_i and p_{i+1}, which is as yet not known. The value of p in Equation (53) can be found iteratively using first Equation (54) below, followed by Equation (55) and finally Equation (53), and repeating Equations (55) and (53) if necessary.

$$p_i = p_{i-1} + \frac{1}{8 N_p p_{i-1}{}^3 e^{-2 p_{i-1}{}^2}} \tag{54}$$

$$p = \frac{p_{i-1} + p_i}{2} \tag{55}$$

Repeating for $i = 2$ to N_p results in a list of N_p elements spaced such that the distribution matches Equation (14). Best results are achieved if the list of values is randomly rearranged (once) before use. The list of values can then be used to assign values of p for generating rays using Equation (12).

Suppose we want to generate a set of $N_r = 4096$ rays. We could use 128 values of α and 32 values of γ and δ corresponding to dodecahedron and icosahedron vertices. We want to have sets of 8 rays with the same value of p for offset calculations, four with γ and four with $\gamma + \pi$, as will be explained later. So, we can make a list of $N_p = 512$ values of p. We would generate the ray set as follows:

1. Select 32 values of α equally spaced between 0 and $\pi/2$.
2. Select 16 values of γ and δ (omitting vertices which are opposite each other, since $\gamma + \pi$ will provide these points).
3. For each of the 512 combinations of α, γ, and δ selected in steps (1) and (2), select a value of p from the list of p values and use Equation (12) to generate 4 rays with α, $\alpha + \pi/2$, $\alpha + \pi$, and $\alpha + 3\pi/2$ and γ, and 4 rays with α, $\alpha + \pi/2$, $\alpha + \pi$, and $\alpha + 3\pi/2$ and $\gamma + \pi$.

This provides us with a deterministic and repeatable selection of rays which has the property that the ray density is proportional to the Gaussian

Representation of Gaussian Beams Using Rays 35

field. The next step is to propagate these rays and compute the output field distribution.

Phase Offsets

The angular spectrum calculation will produce the correct Gaussian field distribution if plane waves corresponding to the far-field angular distribution are summed with the phases matching at the center of the beam waist. However, most rays generated by Equation (12) will originate with non-zero values of x, y, L, and M, so in order for the associated plane wave to have zero phase at the origin (which is the center of the beam waist), we need to add a phase offset

$$\phi = \frac{2\pi n}{\lambda}(xL + yM). \tag{56}$$

The phase offset given by Equation (56) is shown as a path offset in Figure 11(a). If we apply this path offset to each ray and then try to reconstruct the field at the input surface, each ray will have zero phase at the center of the waist (point O) and we will obtain the correct Gaussian field. However, if the rays propagate through a lens, we find that a different path offset is needed to make the associated plane waves align in phase at the center of the output waist. An example is shown in Figure 11(b), where the input waist and the output waist are both a distance f from a lens of focal length f. From Fermat's principle, if we consider Figure 11(b) to be half of an imaging system (place a mirror at plane BD), the ray path OAB and the ray path OCD must have the same optical path length. Now considering the ray EFD, if we apply Fermat's theorem in the reverse direction, with the object at point D, we can show that the optical path length of EFD is shorter than OCD by the amount $-xL$. The coordinate of ray EFD at the output is $x_{out} = Lf$, and the direction cosine is $L_{out} = -x/f$, so a path correction of $-xL$ is needed at the output in order for ray EFD have the same phase as ray OAB at the point B at the center of the output waist.

But, the ray *EFD* is shorter than ray *OAB* by exactly the amount $-xL$, so no additional path offset is necessary.

We therefore have two cases: (1) no lens, where the input and output rays both have coordinates x and L at the beam waist and a path correction of xL is needed, and (2) with a lens where the input coordinates are x and L, the output coordinates are Lf and $-x/f$ and zero path correction is needed. Some simple algebra tells us that applying half the phase offset of Equation (56) at the input and half at the output meets the phase requirements for both case (1) and case (2), and in fact we find that this approach results in the correct phase offsets for any aberration-free lens, or any combination of lenses.

There is one significant difficulty though, in that we may not be able to determine the location of the beam waist at the output if there are aberrations present or if the beam is generally astigmatic, and if we compute Equation (56) at different z locations we will get different results because the values of x and y will change. We need a way to compute the output phase offset which does not depend on knowing the output waist position. We can use Equation (43) to substitute for x and y in Equation (56), which results in the output phase offset (where the phase offset is being calculated for ray 1, and L_2 and M_2 are the direction cosines of a ray with $\alpha + \pi/2$ and the same values of p, γ, and δ as ray 1)

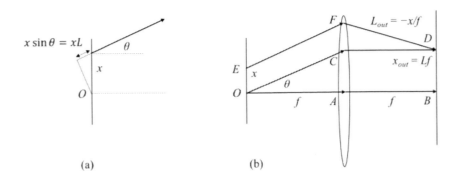

Figure 11. (a) The path correction required for a ray launched with nonzero values of x and L, to achieve zero phase at the center of the beam waist O, is equal to xL. (b) If the beam passes through a lens, with the input and output waists one focal length from the lens, no path correction is needed for ray *EFD* to have zero phase offset at the point B.

Representation of Gaussian Beams Using Rays

$$\phi_{out} = -\frac{L_1 L_2}{\theta_{0x}^2} - \frac{M_1 M_2}{\theta_{0y}^2} \, . \tag{57}$$

Equation (57) contains only ray direction cosines, so we do not need to know the output waist position to calculate the output phase offsets. We do need however θ_{0x} and θ_{0y}, so we should evaluate Equation (57) in the coordinates of the far-field divergence, the orientation of which is determined by Equation (42) but summing over all propagated rays instead of just four.

$$\varphi_D = \frac{1}{2}\tan^{-1}\left(\frac{2\left(\sum L_j M_j\right)}{\sum L_j^2 - \sum M_j^2}\right) \tag{58}$$

Then, after converting all ray direction cosines into the coordinates of the far-field divergence, the values of θ_{0x} and θ_{0y} can be determined using Equation (41), but summing over all propagated rays.

$$\theta_{0x}^2 = \frac{2}{N_r}\sum_{j=1}^{N_r}\frac{L_j^2}{p_j^2}$$

$$\theta_{0y}^2 = \frac{2}{N_r}\sum_{j=1}^{N_r}\frac{M_j^2}{p_j^2} \tag{59}$$

Now it is clear from Equation (57) why we need to propagate the rays in sets of eight with the same value of p, so that we have a ray with $\alpha + \pi/2$ available for each propagated ray, to enable calculation of the output phase offset.

The input phase offset is half of Equation (56), and we can use Equation (12) to get an alternate expression for the input phase offset.

$$\phi_{in} = \frac{\pi n}{\lambda}(xL + yM) = p^2[\sin(\alpha + \gamma)\cos(\alpha + \gamma) - \sin\alpha\cos\alpha] \tag{60}$$

Field Calculation

We are now in a position to calculate the output field profile. This is done by simply summing the plane wave field contributions from each ray. Since the rays have a distribution of angles which matches the Gaussian beam far field profile, we can assign unity weight to each ray. Then, the field distribution at the output is given by

$$E(x,y) = \frac{1}{N_r}\sum_{j=1}^{N_r} e^{i\left(\frac{2\pi}{\lambda}[n(x-x_j)L_j+n(y-y_j)M_j+OPL_j]+\phi_{inj}+\phi_{outj}\right)} \quad (61)$$

where n is the refractive index at the output surface, OPL_j is the optical path length (integrated propagation distance times refractive index) of the j'th ray from the input surface to the output surface, and x_j, y_j, L_j, and M_j are in the coordinates of the output surface (without the central ray coordinates subtracted); however note that when using Equations (57-59) to compute ϕ_{out}, the ray coordinates with the central ray subtracted are used. When evaluating Equation (61) over a rectangular grid of $d{\times}d$ points in x and y, considerable savings in computation time can be achieved by making use of the trigonometric identities

$$\cos(a + b) = \cos a \cos b - \sin a \sin b$$
$$\sin(a + b) = \sin a \cos b + \sin a \cos b. \quad (62)$$

Letting

$$a = \frac{2\pi}{\lambda}\left[n(x - x_j)L_j\right]$$
$$b = \frac{2\pi}{\lambda}\left[n(y - y_j)M_j + OPL_j\right] + \phi_{inj} + \phi_{outj} \quad (63)$$

the computation of Equation (61) requires only $4d$ trigonometric function calculations per ray, instead of $2d^2$.

Coupling Efficiency

Once the field distribution is known, computing coupling efficiency into a fiber or waveguide mode with mode field distribution $B(x,y)$ is accomplished using the mode overlap integral.

$$\eta = \frac{|\iint E(x,y)B^*(x,y)dxdy|^2}{\iint E(x,y)E^*(x,y)dxdy \iint B(x,y)B^*(x,y)dxdy} \tag{64}$$

If the mode field distribution $B(x,y)$ is a Gaussian mode

$$B(x,y) = e^{-\left(\frac{x^2}{\omega_{0xT}^2}+\frac{y^2}{\omega_{0yT}^2}\right)} \tag{65}$$

the evaluation of the numerator of Equation (64) simplifies considerably.

$$\iint E(x,y)B^*(x,y)dxdy$$

$$= \iint \frac{1}{N_r}\sum_{j=1}^{N_r} e^{i\left(\frac{2\pi}{\lambda}[n(x-x_j)L_j+n(y-y_j)M_j+OPL_j]+\phi_{inj}+\phi_{outj}\right)} e^{-\left(\frac{x^2}{\omega_{0xT}^2}+\frac{y^2}{\omega_{0yT}^2}\right)}dxdy$$

$$= \frac{1}{N_r}\sum_{j=1}^{N_r} e^{i\left(\frac{2\pi}{\lambda}[-nx_jL_j-ny_jM_j+OPL_j]+\phi_{inj}+\phi_{outj}\right)}$$

$$\cdot \iint e^{i\left(\frac{2\pi n}{\lambda}[xL_j+yM_j]\right)} e^{-\left(\frac{x^2}{\omega_{0xT}^2}+\frac{y^2}{\omega_{0yT}^2}\right)}dxdy$$

$$= \frac{\pi\omega_{0xT}\omega_{0yT}}{N_r}$$

$$\cdot \sum_{j=1}^{N_r} e^{i\left(\frac{2\pi}{\lambda}[-nx_jL_j-ny_jM_j+OPL_j]+\phi_{inj}+\phi_{outj}\right)} e^{-\left(\frac{\pi n L_j\omega_{0xT}}{\lambda}\right)^2} e^{-\left(\frac{\pi n M_j\omega_{0yT}}{\lambda}\right)^2} \tag{66}$$

In Equation (66), the double integral evaluates to an explicit formula and therefore the need for integration or even the calculation of the field profile $E(x,y)$ is eliminated. The terms in the denominator of Equation (64) can be evaluated using

$$\iint E(x,y)E^*(x,y)dxdy = \frac{\lambda^2}{\pi n^2\theta_{0x}\theta_{0y}} \tag{67}$$

$$\iint B(x,y)B^*(x,y)dxdy = \frac{\pi\omega_{0xT}\omega_{0yT}}{2} \qquad (68)$$

where θ_{0x} and θ_{0y} in Equation (67) are calculated using Equation (59). We now substitute Equations (66), (67), and (68) into Equation (64) to determine the coupling efficiency of the output field into the Gaussian mode defined by Equation (65). If apertures are present in the system, such that some rays are blocked, Equation (64) still gives the correct answer for the total coupled power, the power lost at the apertures being accounted for by the omission of the contribution of those rays in Equation (66). However, if some rays are blocked, steps must be taken to replace the lost information needed in Equation (57) for the calculation of output phase offsets of other rays. For this, a ray might be propagated at a smaller value of p which will not be blocked, and the resulting values of L and M scaled to provide estimates for L and M of the blocked ray.

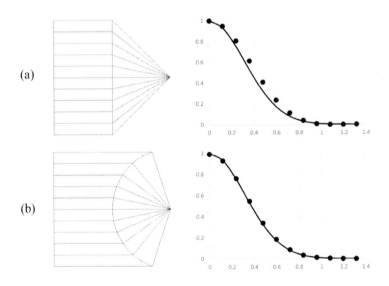

Figure 12. Comparison of intensity calculations using Equation (61) with 4096 rays (markers) with the expected Gaussian shape (solid line) for a tightly focused Gaussian beam with $\omega_0 = 0.64$ μm and $\lambda = 1$ μm, (a) when the beam is focused by a planar lens such as a diffractive lens, and (b) when the beam is focused by a lens meeting the Abbe sine condition. The calculated intensity in (b) is a better match with the expected Gaussian mode profile.

Limitations

It is important to mention again the main limitation of this ray-based method of field calculation and coupling efficiency calculation. As in Wagner and Tomlinson's method [12], the aberrations must occur in the far-field of the output mode. Essentially this means that the ray aberrations must be primarily in x and y (relative to ω_{0x} and ω_{0y}) or in phase, not L and M (relative to θ_{0x} and θ_{0y}). In the case that there are aberrations in L and M, it has been found that more accurate results are obtained if the values of L and M are obtained from paraxial calculations rather than using the aberrated values of L and M. Paraxial values can be obtained by tracing rays with a small value of p and scaling the results to the value of p used for the real ray.

Alternatively, matrix algebra can be used to obtain the paraxial output ray coordinates x', y', L', and M' corresponding to any input ray coordinates x, y, L, and M if the ray transfer matrix $\mathbf{M}$ can be determined for the optical system.

$$
\begin{pmatrix} x' \\ y' \\ L' \\ M' \end{pmatrix} = \begin{bmatrix} m_{11} & m_{12} & m_{13} & m_{14} \\ m_{21} & m_{22} & m_{23} & m_{24} \\ m_{31} & m_{32} & m_{33} & m_{34} \\ m_{41} & m_{42} & m_{43} & m_{44} \end{bmatrix} \begin{pmatrix} x \\ y \\ L \\ M \end{pmatrix}
\tag{69}
$$

If we trace four rays, we have 16 equations and 16 unknowns so we can solve for the elements of M. A convenient choice of rays would be the rays defined by Equation (12) and Table 1, with a small value of p and $\gamma = \delta = 0$. The matrix $\mathbf{M}$ is then given by

$$
[\mathbf{M}] = \frac{1}{2p} \begin{bmatrix} x_1' & x_2' & x_3' & x_4' \\ y_1' & y_2' & y_3' & y_4' \\ L_1' & L_2' & L_3' & L_4' \\ M_1' & M_2' & M_3' & M_4' \end{bmatrix} \begin{bmatrix} \dfrac{1}{\omega_{0x}} & 0 & 0 & \dfrac{1}{\theta_{0y}} \\ 0 & \dfrac{1}{\omega_{0y}} & -\dfrac{1}{\theta_{0x}} & 0 \\ \dfrac{1}{\omega_{0x}} & 0 & 0 & -\dfrac{1}{\theta_{0y}} \\ 0 & -\dfrac{1}{\omega_{0y}} & -\dfrac{1}{\theta_{0x}} & 0 \end{bmatrix} .
\tag{70}
$$

42   *Paul D. Colbourne*

When paraxial values of L and M are used, some simplification of the calculations can be made. Equations (41) and (42) can be used instead of Equations (58) and (59), and there is no need for special measures to calculate the direction cosines of blocked rays for use in Equation (57).

Direction Cosines vs. Ray Slopes

We can now make an additional observation in support of the choice of defining the ray angles in Equation (12) as direction cosines instead of ray slopes. If a very small waist size is specified, less than the wavelength, we see that the calculated field matches better the expected Gaussian field when the ray definitions of Equation (12) specify the ray angles in terms of their direction cosines (Figure 12(b)), than if we were to rewrite Equation (12) to specify the ray angles in terms of the ray slopes (Figure 12(a)). This has an additional consequence when it comes to real lens systems which focus or collimate highly divergent beams: such lens systems should meet the Abbe sine condition [13]. A focusing lens meeting the Abbe sine condition (such as a microscope objective) will transform the ray set of a collimated (low divergence) Gaussian beam into a ray set corresponding to Equation (12) (illustrated in Figure 12(b)), whereas a flat diffractive lens (which does not meet the Abbe sine condition) used as a focusing lens will produce a ray set corresponding to rays defined by ray slopes (as shown in Figure 12(a)), and the angular spectrum field calculations show that the lens meeting the Abbe sine condition produces the better focused spot.

CONCLUSION

This chapter showed how skew rays can be used to represent Gaussian beams in generally astigmatic optical systems, and can be used to calculate the properties of the propagating generally astigmatic Gaussian beam in a straightforward manner. Furthermore, skew rays can be used to optimize an optical system to minimize aberrations and obtain the correct Gaussian

Representation of Gaussian Beams Using Rays 43

beam focus. The optimization method requires very little computation beyond that of a conventional ray optimization, requires no explicit calculation of the properties of the propagating Gaussian beam, and works even in optical systems with general astigmatism. Further, a ray set may be defined which has a ray density equal to the Gaussian field amplitude or intensity distribution at all points. Such a ray set can be used for Monte Carlo analysis of stray light, or it can be used to compute the output field profile or the coupling efficiency into a fiber or waveguide mode. A simplified calculation was presented for calculating the coupling efficiency when the mode to be coupled into has a Gaussian profile.

REFERENCES

[1] Steier, William H. (1966). "The ray packet equivalent of a Gaussian light beam," *Applied Optics*, 5, 1229-1233.

[2] Arnaud, Jacques. (1985). "Representation of Gaussian beams by complex rays," *Applied Optics*, 24, 538-543.

[3] Herloski, Robert., Marshall, Sidney. & Antos, Ronald. (1983). "Gaussian beam ray-equivalent modeling and optical design," *Applied Optics*, 22, 1168-1174.

[4] Greynolds, Alan W. (1985). "Propagation of general astigmatic Gaussian beams along skew ray paths," *Proceedings of the SPIE*, *560*, 33–50.

[5] Arnaud, J. A. (1970). "Nonorthogonal Optical Waveguides and Resonators," *Bell Systems Technical Journal*, Nov. 1970, 2311-2348.

[6] Lü, Baida., Feng, Guoying. & Cai, Bangwei. (1993). "Complex ray representation of the astigmatic Gaussian beam propagation," *Optical and Quantum Electronics*, 25, 275-284.

[7] Colbourne, Paul D. (2014). "Generally astigmatic Gaussian beam representation and optimization using skew rays," *Proceedings of the SPIE*, *9293*, 92931S.

44   *Paul D. Colbourne*

[8] Pakhomov, I. I. & Tsibulya, A. B. (1988). "Computational Methods for Laser Optical Systems Design", *Journal of Russian Laser Research*, *9*, 321-430.

[9] Colbourne, Paul D. (2017). "Ray optimization of Gaussian beams," *2017 Photonics North (PN)*, Ottawa, ON, 2017, 1-1.

[10] Box, G. E. P. & Muller, M. E. (1958). "A Note on the Generation of Random Normal Deviates," *The Annals of Mathematical Statistics*, *29*, 610–611.

[11] Arnaud, J. A. & Kogelnik, H. (1969). "Gaussian Light Beams with General Astigmatism," *Applied Optics*, *8*, 1687-1693.

[12] Wagner, R. E. & Tomlinson, W. J. (1982). "Coupling efficiency of optics in single-mode fiber components," *Applied Optics*, *15*, 2671-2688.

[13] Mansuripur, Masud. (1998). "Abbe's Sine Condition," *Optics and Photonics News*, February 1998, 56-60.

BIOGRAPHICAL SKETCH

Paul D. Colbourne

Affiliation: Lumentum Ottawa Inc., Ottawa, Ontario, Canada

Education: Ph.D. from McMaster University, Hamilton, Ontario, Canada.

Business Address: 61 Bill Leathem Drive, Ottawa, ON, Canada K2J 0P7

Research and Professional Experience: Dr. Paul Colbourne is Director, Optical Switching Technology Advanced Research at Lumentum (and formerly JDS Uniphase) in Ottawa, Ontario, Canada, where he has been designing optical components for fiber optic communication systems for over 25 years. He has been an inventor on more than 50 US patents and has authored more than 30 papers and conference presentations.

Representation of Gaussian Beams Using Rays   45

Professional Appointments:

Honors: Senior Member of the Optical Society (OSA).

Publications from the Last 3 Years:

Colbourne, Paul D., McLaughlin, Sheldon., Murley, Chester., Gaudet, Simon., & Burke, Dan. (2018). "Contentionless Twin 8x24 WSS with Low Insertion Loss," *Optical Fiber Communication Conference Postdeadline Papers*, OSA Technical Digest (online) (Optical Society of America, 2018), paper Th4A.1.

Colbourne, Paul D. (2017). "Ray optimization of Gaussian beams," *2017 Photonics North (PN)*, Ottawa, ON, 2017, 1-1.

Marom, Dan M., Colbourne, Paul D., D'Errico, Antonio., Fontaine, Nicolas K., Ikuma, Yuichiro., Proietti, Roberto., Zong, Liangjia., Rivas-Moscoso, José M., & Tomkos, Ioannis. (2017). "Survey of Photonic Switching Architectures and Technologies in Support of Spatially and Spectrally Flexible Optical Networking," *Journal of Optical Communications and Networking*, *9*, 1-26.

In: Advances in Engineering Research
Editor: Victoria M. Petrova

ISBN: 978-1-53616-092-5
© 2019 Nova Science Publishers, Inc.

Chapter 2

NOVEL I/O-LDC CONTROL BASED ON ANFIS FOR VARIABLE SPEED WIND-TURBINE SYSTEM

Fayssal Amrane[1,] and Azeddine Chaiba[1,2]*

[1]Automatic Laboratory of Setif (LAS), Department of Electrical Engineering, University of Setif 1, Setif, Algeria

[2]Department of industrial Engineering, University of Khenchela, Khenchela, Algeria

ABSTRACT

This chapter presents an enhanced Input/Output-Linearizing and Decoupling Control (I/OLDC) in variable speed for Wind Energy Conversion System (WECS), which is based on ANFIS (Adaptive Neuro-Fuzzy Inference System). ANFIS is a combination of two soft-computing methods of ANN (Artificial Neural Network) and fuzzy logic. Fuzzy logic has the ability to change the qualitative aspects of human knowledge and insights into the process of precise quantitative analysis. The improved I/OLDC is applied in variable speed for WECS using Doubly Fed Induction Generator (DFIG) fed by back-to-back two-phase voltage source inverter (VSI). The proposed control technique is used in

---

[*] Corresponding Author's Email: amrane_fayssal@live.fr.

order to control the stator active and reactive powers of DFIG by the means of MPPT (Maximum Power Point Tracking) strategy. By using the feedback linearization, the control algorithm is established. In order to improve the tracking of stator active and reactive power references, ANFIS is used after the comparison between measured and reference of stator active and reactive powers respectively. The proposed control is used to overcome the drawbacks of the classical control such as PI controllers in terms of; overshoot, response time, power error and the reference tracking. Finally, simulation results demonstrate that the proposed control using ANFIS provides improved dynamic responses and perfect decoupled control of the wind turbine has driven DFIG with high performances (good reference tracking, short response time and neglected power error) in steady and transient states.

Keywords: ANFIS (Adaptive Neuro-Fuzzy Inference System), DFIG (Doubly Fed Induction Generator), WECS (Wind Energy Conversion System), PI (Proportional Integral), VSI (Voltage Source Inverter), I/OLDC (Input/Output Linearizing & Decoupling Control), ANN (Artificial Neural Network) & FL (Fuzzy Logic)

1. INTRODUCTION

During the past decade, the installed wind power capacity in the world has been increasing more than 30% (Fayssal Amrane 2016a). Doubly Fed Induction Generator (DFIG) based wind turbines (WTs) have many advantages over the fixed speed induction generators or variable speed synchronous generators with full-scale power converters, including variable speed operation for maximum power tracking, decoupled active and reactive power control, lower converter cost, and reduced power loss (Fayssal Amrane 2016b; M. Jujawa 2003; J. Yao 2008). The electronic interface dealing with wind generator control is basically a back-to-back two-level converter; however this kind of power converter is seldom used for high power applications (J. Guo 2008).

Nowadays, since DFIG-based WTs (Figure 1) are mainly installed in remote and rural areas (Akbar Tohidi 2016). In literature (Jafar Mohammadi 2014) vector control is the most popular method used in the

DFIG-based wind turbines (WTs). In most applications, the proportional-integral (PI) controller based Power Field Orieted Control (FOC) scheme is used to control DFIG in wind energy conversion systems (WECSs) (Roberto Cardenas 2013; Etienne Tremblay 2011). Although this control scheme is easy to implement, it has some drawbacks. One of the most important drawbacks of this control scheme is that the performance of the PVC (*known also by Indirect Power Control 'IDPC'*) scheme largely depends on the tuning of the PI controller's parameters (K_p and K_i). Another drawback of this controller is that its performance also depends on the accuracy of the machine parameters and on grid voltage conditions such as harmonic level, distortion, etc. (Etienne Tremblay 2011). However, the switching frequency of the converter is still affected significantly by active and reactive power variations and operating speed. Although, DFIG control using Input-Output Feedback Linearization method can operate below and above synchronous speed (H. Stemmler 1993). This algorithm control offers high performance dynamics with decoupling of stator active and recative power in transient and steady states compared with power vector control.

In order to control the IGBTs of back-to-back converter; Space Vector modulation (SVM) will present the best solution (Fayssal Amrane 2016b) compared to Pulse Width Modulation (PWM) in order to minimize the harmonic and keeping the fixed switching frequency of AC-DC-AC converters (*especially in Rotor Side Converter 'RSC'*).

To extract the maximum power despite sudden variations in wind speed, the maximum power point tracking (MPPT) strategy (Yu Zou 2013; Shenghu LI 2013; Baike Shen 2009; Z. -S. Zhang 2012 and N. K. Swami Naidu 2016) is proposed, the stator active power is extracted from wind power and stator reactive power is maintained at zero level to ensure unity power (PF≈1).

Actually, the active and reactive powers of the DFIG are controlled by regulating the amplitudes and phase angles of the stator and rotor fluxes. The application of the classical input/output nonlinear control based on PI regulator generates some drawbacks or limitation of the performances (*power error, tracking power, ...etc.*).

It noted that in recent years, the application of Adaptive Neuro-Fuzzy System (ANFIS) is extended especially in the WECS in order to remove or mitigate the drawbacks of conventional controllers. In this context, intelligent algorithms such as Neuro-fuzzy Control (NFC) are used instead the PI controllers.

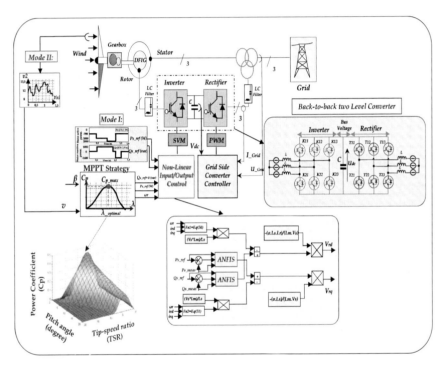

Figure 1. Schematic diagram of wind turbine system with a DFIG.

Furthermore fuzzy-neural techniques have been proposed as a robust control for electrical drives (Azeddine Chaiba 2012; Cetin Elmas 2008 and Muammer Gökbulut 2006). The ANFIS architecture has well known advantages of modeling a highly Non-Linear system, as it combines the capability of fuzzy reasoning in handling the uncertainties and capability of an artificial neural network (ANN) in learning from processes (K. Kouzi 2008 and Rodrigo Marchi 2014). The main purpose of using the ANFIS (*Adaptive Neuro-Fuzzy Inference Systems*) approach is to automatically deliver the fuzzy system by using neural network methods. A combination of the strengths of Fuzzy Logic controllers and Neural Networks creates systems capable of controlling complex systems and adaptively learning to optimize control parameters (K. Kouzi 2008 and Rodrigo Marchi 2014). These advantages justify the necessity of applying this kind of system for the DFIG used in WECS.

The main contribution of this chapter is the validation of the proposed control using NFC based on input/output linearizing control to provide improved decoupled performance and parameter, also to control independently the stator active and reactive powers of DFIG-wind turbine. ANFIS used to improve powers tracking and overcome the drawbacks of the classical nonlinear control based on PI regulators such as; long response time, remarkable overshoot, and big power error.

This chapter is organized as follows; firstly the mathematical model of ANFIS under four (04) layers is presented in details in section II. The main idea of proposed chapter is explained in section III. The mathematical model of wind turbine and MPPT is illustrated in details in section IV. The DFIG model is presented in section V. In section VI input-output linearizing control of the DFIG is presented. This section is divided into two parts as follows: general concept of the input-output linearizing control & control of rotor side converter (RSC). In section VII illustrated the grid side converter (GSC) and dc-link voltage control. The RSC is developed in section VIII, in this section Space Vector modulation (SVM) is used instead PWM (Pulse Widh Modulation) and more explication are presented. Operating principle of DFIG is illustrated in section IX in order to explain the four quadrants of DFIG operating under Sub and Super-synchronous modes. In section X, computer simulation results are shown and discussed for conventional and improved algorithm control in transient and steady states. Finally, the reported work is concluded.

2. ADAPTIVE NEURO-FUZZY INFERENCE SYSTEM (ANFIS)

The block diagram of the neuro-fuzzy controller (NFC) system[1] is shown in Figure 2. The NFC controller is composed of an on-line learning algorithm with a neuro-fuzzy network. The neuro-fuzzy network is trained using an on-line learning algorithm. The NFC has two inputs, the stator active error e_{Ps} and the derivative of stator active error $\dot{e}_{Ps}$. The output is rotor direct voltage V_{dr}. For the NFC of stator recative power Q_s is similar with P_s controller (Fayssal Amrane 2015). For the NFC, a four layer NN as shown in Figure 4 is used. Layers I–IV represents the inputs of the network, the membership functions, the fuzzy rule base and the outputs of the network, respectively (Fayssal Amrane 2016c). Knowing that, the NFC parameters are illustrated in details in Table 1. The training error is depicted in Figure 3, which equals to 0.05; presents very good precision (Fayssal Amrane 2016c).

2.1. Layer I: Input Layer

Inputs and outputs of nodes in this layer are represented as:

$$net_1^I = e_{Ps}(t).y_1^I = f_1^I(net_1^I) = net_1^I = e_{Ps}(t) \tag{1}$$

$$net_2^I = \dot{e}_{Ps}(t).y_2^I = f_2^I(net_2^I) = net_2^I = \dot{e}_{Ps}(t) \tag{2}$$

Where e_{Ps} and $\dot{e}_{Ps}$ are inputs y_1^I and y_2^I are outputs of the input layer. In this layer, the weights are unity and fixed.

---

[1] In this chapter, we used Neuro fuzzy controller (NFC) & Adaptive Neuro-Fuzzy System (ANFIS); this later is more generalized than NFC.

Novel I/O-LDC Control Based on ANFIS ... 53

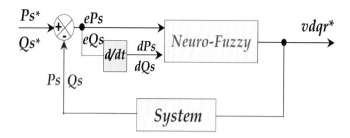

Figure 2. Block diagram of the neuro-fuzzy controller under Matlab/Simulink®.

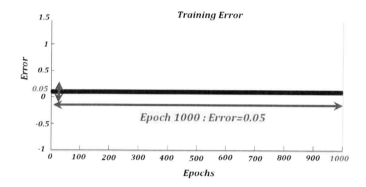

Figure 3. Simulation results of training error (0.05) using 1000 epochs.

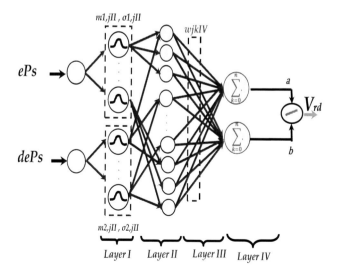

Figure 4. Schematic diagram of the neuro-fuzzy network.

Fayssal Amrane and Azeddine Chaiba

Table 1. The parameters of the proposed Neuro-fuzzy controller

Type:	Takagi Sugeno.
Number of iteration:	500.
Error tolerance:	$5*10^{-3}$.
Epochs:	1000.
Number of membership function:	7.
Number of hidden layer neuron:	14.

2.2. Layer II: Membership Layer

In this layer, each node performs a fuzzy set and the Gaussian function is adopted as a membership function.

$$net_{1,j}^{II} = -\frac{(x_{1,j}^{II} - m_{1,j}^{II})^2}{(\sigma_{1,j}^{II})^2} . y_{1,j}^{II} = f_{1,j}^{II}(net_{1,j}^{II}) = \exp(net_{1,j}^{II}) \tag{3}$$

$$net_{2,k}^{II} = -\frac{(x_{2,k}^{II} - m_{2,k}^{II})^2}{(\sigma_{2,k}^{II})^2} . y_{2,k}^{II} = f_{2,k}^{II}(net_{2,k}^{II}) = \exp(net_{2,k}^{II}) \tag{4}$$

Where: $m_{1,j}^{II}, m_{2,k}^{II}$ and $\sigma_{1,j}^{II}, \sigma_{2,k}^{II}$ are respectively, the mean and the standard deviation of the Gaussian function. There are j+k nodes in this layer.

2.3. Layer III: Rule Layer

This layer includes the rule base used in the fuzzy logic control (*FLC*). Each node in this layer multiplies the input signals and outputs the result of product (Azeddine Chaiba 2012).

$$net_{j,k}^{III} = (x_{1,j}^{III} . x_{2,k}^{III}) . y_{jk}^{III} = f_{jk}^{III}(net_{jk}^{III}) = net_{jk}^{III} \tag{5}$$

Where the values of link weights between the membership layer and rule base layer are unity.

2.4. Layer IV: Output Layer

This layer represents the inference and defuzzification used in the *FLC*. For defuzzification, the center of area method is used; therefore the following form can be obtained:

$$a = \sum_j \sum_k w_{jk}^{IV} . y_{jk}^{III} , b = \sum_j \sum_k y_{jk}^{III} \tag{6}$$

$$net_0^{IV} = \frac{a}{b} . y_0^{IV} = f_0^{IV} . (net_0^{IV}) = \frac{a}{b} \tag{7}$$

Where y_{jk}^{III} is the output of the rule layer; a and b are the numerator and the denominator of the function used in the center of area method, w_{jk}^{IV} is the center of the output membership functions used in the FLC, respectively.

The aim of the learning algorithm is to adjust the weights of $w_{jk}^{IV}, m_{1,j}^{II}, m_{2,k}^{II}, \sigma_{1,j}^{II}, \sigma_{2,k}^{II}$ The on-line learning algorithm is a gradient descent search algorithm in the space of network parameters. The error expression for the input of *Layer IV*:

$$\delta_0^{IV} = -\frac{\partial e_{Ps}(t).\dot{e}_{Ps}(t)}{\partial y_0^{IV}} . \frac{\partial y_0^{IV}}{\partial net_0^{IV}} = \mu_5 e_{Ps} \tag{8}$$

Where μ_5 is the learning-rate for w_{jk}^{IV} and it can be shown in following equation.

Therefore, the changing of w_{jk}^{IV} is written as:

$$\Delta w_{jk}^{IV} = -\frac{\partial e_{Ps}(t).\dot{e}_{Ps}(t)}{\partial net_0^{IV}} . \frac{\partial net_0^{IV}}{\partial a} . \frac{\partial a}{\partial w_{jk}^{IV}} = \frac{1}{b} \delta_0^{IV} . y_{jk}^{III} \tag{9}$$

Since the weights in the rule layer are unified, only the approximated error term needs to be calculated and propagated by the following equation:

$$\delta_{jk}^{III} = -\frac{\partial e_{Ps}(t).\dot{e}_{Ps}(t)}{\partial net_0^{IV}} \cdot \frac{\partial net_0^{IV}}{\partial y_{1j}^{III}} \cdot \frac{\partial y_{1j}^{III}}{\partial net_{jk}^{III}} = \frac{1}{b}\delta_0^{IV}.(w_{jk}^{IV} - y_0^{IV}) \tag{10}$$

The error received from *Layer III* is computed as:

$$\delta_{jk}^{II} = \sum_k \left[\left(-\frac{\partial e_{Ps}(t).\dot{e}_{Ps}(t)}{\partial net_{jk}^{III}} \right) \frac{\partial net_{jk}^{III}}{\partial y_{1j}^{II}} \cdot \frac{\partial y_{1j}^{II}}{\partial net_{1,j}^{II}} \right] = \sum_k \delta_{jk}^{III}.y_{jk}^{III} \tag{11}$$

$$\delta_{2k}^{II} = \sum_j \left[\left(-\frac{\partial e_{Ps}(t).\dot{e}_{Ps}(t)}{\partial net_{jk}^{III}} \right) \frac{\partial net_{jk}^{III}}{\partial y_{2j}^{II}} \cdot \frac{\partial y_{2j}^{II}}{\partial net_{2,j}^{III}} \right] = \sum_k \delta_{jk}^{III}.y_{jk}^{III} \tag{12}$$

The updated laws of: $m_{1,j}^{II}, m_{2,k}^{II}$ and $\sigma_{1,j}^{II}, \sigma_{2,k}^{II}$ also can be obtained by the gradient decent search algorithm:

$$\Delta m_{1,j}^{II} = -\frac{\partial e_{Ps}(t).\dot{e}_{Ps}(t)}{\partial net_{1,j}^{II}} \cdot \frac{\partial net_{1,j}^{II}}{\partial m_{1,j}^{II}} = \mu_4.\delta_{1,j}^{II}.\frac{2.(x_{1,j}^{II} - m_{1,j}^{II})}{(\sigma_{1,j}^{II})^2} \tag{13}$$

$$\Delta m_{2,k}^{II} = -\frac{\partial e_{Ps}(t).\dot{e}_{Ps}(t)}{\partial net_{2,k}^{II}} \cdot \frac{\partial net_{2,k}^{II}}{\partial m_{2,k}^{II}} = \mu_3.\delta_{2,k}^{II}.\frac{2.(x_{2,k}^{II} - m_{2,k}^{II})}{(\sigma_{2,k}^{II})^2} \tag{14}$$

$$\Delta \sigma_{1,j}^{II} = -\frac{\partial e_{Ps}(t).\dot{e}_{Ps}(t)}{\partial net_{1,j}^{II}} \cdot \frac{\partial net_{1,j}^{II}}{\partial \sigma_{1,j}^{II}} = \mu_2.\delta_{1,j}^{II}.\frac{2.(x_{1,j}^{II} - m_{1,j}^{II})^2}{(\sigma_{2,k}^{II})^3} \tag{15}$$

Where: μ_4, μ_3, μ_2 and μ_1 are the learning-rate parameters of the mean and the standard deviation of the Gaussian functions.

3. APPLICATION OF ANFIS IN WECS

In this section, we invstigate the ANFIS instead PI in order to control stator active and recative powers (P_s & Q_s) of DFIG as shown in Figure 5.

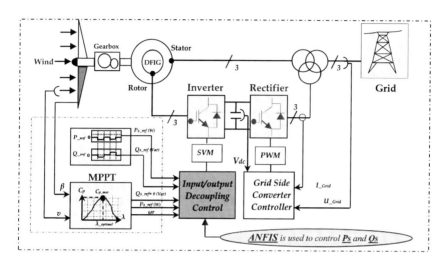

Figure 5. Schematic diagram of wind turbine system using NFC in RSC.

4. WIND TURBINE MATHEMATICAL MODEL

The wind turbine input power usually is (Fayssal Amrane 2017a):

$$Pv = \frac{1}{2}.\rho.Sw.V^3 \qquad (16)$$

Where ρ is air density; S_w is wind turbine blades swept area and V is wind speed.
The output mechanical power of wind turbine is:

$$Pm = Cp.Pv = \frac{1}{2}.Cp.\rho.Sw.V^3 \qquad (17)$$

Where Cp represents power coefficient, λ is tip speed ratio and β the blade pitch angle, λ is given by:

$$\lambda = \frac{R.\Omega t}{v} \tag{18}$$

Where R is blade radius, Ωt is angular speed of the turbine. Cp can be described as (Fayssal Amrane 2017a):

$$Cp = (0.5 - 0.0167.(\beta - 2)).\sin\left[\frac{\pi.(\lambda + 0.1)}{18.5 - 0.3.(\beta - 2)}\right] - 0.00184.(\lambda - 3).(\beta - 2) \tag{19}$$

Knowing that the schematic block of wind turbine is illustrated in Figure 6 and the Theorical maximum power coefficient is presented in Figure 7.

4.1. Maximum Power Point Tracking (MPPT) Strategy

The MPPT strategy without wind speed measurement is illustrates in Figure 8; "So the P_{s_ref} (reference stator power) is calculated by the product of the MPPT's output (the mechanical speed Ω_{mec}) by the electromagnetic torque T_{em_ref}"; so $P_{s\_ref} = \Omega_{mec}* T_{em\_ref}$). Using Matlab/Simulink® to show the behavior of the Power coefficient (C_p) under different pitch angles ($B°$) as shown in Figure 9, it is clear the zero pitch angle ($B° = 0°$) offered the maximum C_p which correspond to Optimal Tip Speed Ratio (TSR). Figure 10 illustrates the three dimensions (3D) of C_p versus TSR and pitch angles ($B°$) respectively.

The main aim of the MPPT strategy; is to adapt the speed of the turbine to the wind speed, in order to maximize the converted power, this will improve its energy efficiency and its integration with the electrical grid. In this chapter, two wind speed profiles are proposed (step wind speed and random wind speed), as shown in Figure 11-(illustrates the simulation results of random wind speed). From the simulation results, it can be seen from Figure 11 that the maximum value of C_p (C_{p_max} = 0.4785) is achieved for $B° = 0°$ & for λ_{Opt} = 8.098. This point corresponds at the maximum power point tracking (MPPT) (Fayssal Amrane 2017a).

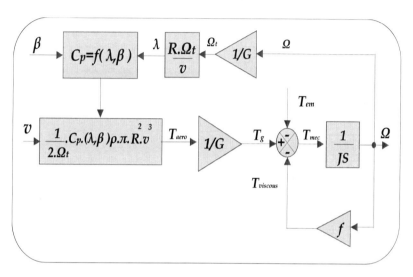

Figure 6. Schematic block of wind turbine.

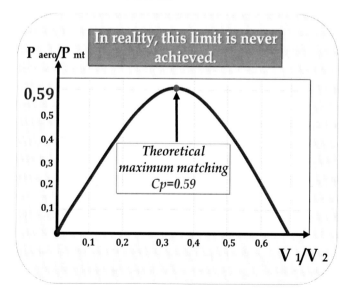

Figure 7. Theorical maximum power coefficient (Cp = (16/27) ≈ 0.59).

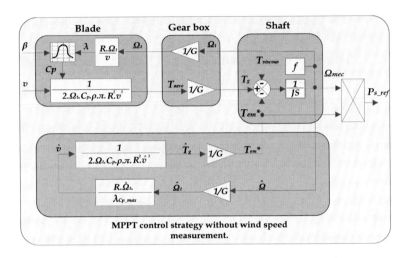

Figure 8. MPPT control strategy without wind speed measurement.

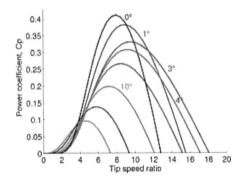

Figure 9. Cp under different pitch angles (B°).

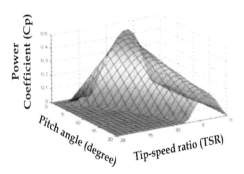

Figure 10. 3D Power coefficient versus Tip speed ratio (TSR) and Pitch angle degree (B°).

Novel I/O-LDC Control Based on ANFIS ... 61

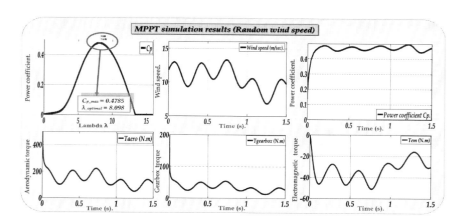

Figure 11. Simulation results of MPPT strategy: (C_p versus λ "or TSR", Wind speed versus time, C_p versus time) T_{aero}, $T_{gearbox}$ & T_{em} using random wind speed profile.

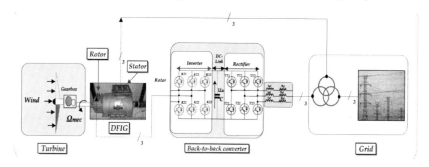

Figure 12. Main circuit topology of a back-to-back SVM converter for DFIG.

After the simulation of the wind turbine model using proposed wind profile, we test the robustness of the MPPT algorithm; we have as results the C_p versus time for proposed wind speed profile; this latter achieved the maximum value mentioned in Figure 11 (C_{p_max} = 0.4785) despite the sudden variation of the wind speed. T_{aero}, $T_{gearbox}$ and T_{em} are depicts respectively in Figure 11 (at the bottom) for proposed wind speed. Knowing that, T_{em} takes negative value (*Generator mode*). Figure 12 illustrates the full circuit topology of a DFIG system with a back-to-back PWM converter, which is composed of a GSC (*Grid Side Converter*), a RSC (*Rotor Side Converter*) and a DC-link capacitor.

5. Mathematical Model of DFIG

The general electrical state model of the DFIG obtained using Park transformation is given by the following equations (G.Abad J.Lopez 2011; Azeddine Chaiba 2013):

Stator and rotor voltages:

$$V_{sd} = Rs.isd + \frac{d}{dt}\phi sd - \omega s.\phi sq \tag{20}$$

$$V_{sq} = Rs.isq + \frac{d}{dt}\phi sq - \omega s.\phi sd \tag{21}$$

$$V_{rd} = Rr.ird + \frac{d}{dt}\phi rd - (\omega s - \omega_r).\phi rq \tag{22}$$

$$V_{rq} = Rr.irq + \frac{d}{dt}\phi rq - (\omega s - \omega_r).\phi rd \tag{23}$$

Stator and rotor fluxes:

$$\phi sd = Ls.isd + Lm.ird \tag{24}$$

$$\phi sq = Ls.isq + Lm.irq \tag{25}$$

$$\phi rd = Lr.ird + Lm.isd \tag{26}$$

$$\phi rq = Lr.irq + Lm.isq \tag{27}$$

The electromagnetic torque is given by:

$$Tem = P.Lm.(ird.isq - irq.isd) \tag{28}$$

$$Tem - Tr = J.\frac{d}{dt}\Omega + f\Omega_{mec} \tag{29}$$

Novel I/O-LDC Control Based on ANFIS ... 63

Where; R_s, R_r, L_r, & L_s are respectively the resistances and the inductances of the stator and the rotor of the *DFIG* & L_m is mutual inductance, σ is leakage factor.

V_{sd}, V_{sq}, V_{rd}, V_{rq}, I_{sd}, I_{sq}, I_{rd}, I_{rq}, Φ_{sd}, Φ_{sq}, Φ_{rd} & Φ_{rq} respectively represent the components along the *d and q* axes of the stator and rotor voltages, currents and flux.

T_{em}, T_r, T_{vis}, T_{aero} and $T_{gearbox}$ present the *"electromagnetic"*, *"load"*, *"viscous"*, *"aerodynamic"* and *"gearbox torques"*. *Jg*, $J_{turbine}$ and *J* are the "generator", "turbine" and "total inertia" in *DFIG's* rotor respectively, Ω_{mec} is the mechanical speed, and *G* is the gain of gear box.

P is number of pole pairs, w_s is the stator pulsation, w_r is the rotor pulsation, w_{slip} is the slip pusation[2] & *f* is the friction coefficient. T_s & T_r are stator & rotor time-constant, and S[3] is the slip $(S = w_s - w/w_s)$.

with:
$$Tr = \frac{Lr}{Rr}; Ts = \frac{Ls}{Rs}; \sigma = 1 - \frac{L_m^2}{Ls.Lr}$$

In this section, the DFIG model can be described by the following state equations in the synchronous reference frame whose axis d is aligned with the stator flux vector as shown in Figure 13, $\Phi_{sd} = \Phi_s$ & $\Phi_{sq} = 0$ (Y.Lei 2006).

By neglecting resistances of the stator phases the stator voltage will be expressed by:

$$Vsd = 0 \text{ and } Vsq = Vs \cong \omega s.\phi s \tag{30}$$

$$\Phi s = Ls.isd + Lm.ird \tag{31}$$

$$0 = Ls.isq + Lm.irq \tag{32}$$

---

[2] Knowing that: wslip = ws-w.
[3] In this case the slip varies between $\pm 0.05 \rightarrow \pm 0.07$ (means the stable zone), refer to Figure 27 (section IX).

From (31) and (32), the equation linking the stator currents to the rotor currents are deduced below:

$$i_{sd} = \frac{\phi_s}{L_s} - \frac{L_m}{L_s}.i_{rd} \tag{33}$$

$$i_{sq} = -\frac{L_m}{L_s}.i_{rq} \tag{34}$$

We lead to an uncoupled power control; where, the transversal component I_{rq} of the rotor current controls the stator active power. The stator reactive power is imposed by the direct component I_{rd} as in shown in Figure 14 (Fayssal Amrane 2016d; Fayssal Amrane 2017b):

$$\begin{cases} P_s = -V_s.\dfrac{L_m}{L_s}.I_{rq} \\ Q_s = \dfrac{V_s^2}{\omega_s.L_s} - V_s.\dfrac{L_m}{L_s}.I_{rd} \end{cases} \tag{35}$$

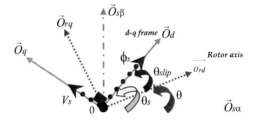

Figure 13. Stator and rotor flux vectors in the synchronous d-q Frame.

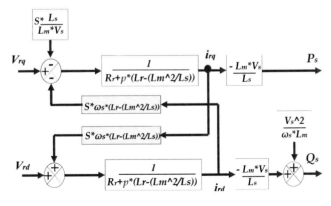

Figure 14. The doubly fed induction generator simplified model.

6. Input/Output Decoupling and Linearizing Control of DFIG

6.1. General Concept of the Input-Output Linearizing Control

Given the system (S. Hu 2010; S. E. Enev 2006 and D.M. Brod 1985) as shown in Figure 15:

$$\begin{cases} \dot{x} = fn(x) + g \times u. \\ \quad y = h(x) \end{cases} \tag{36}$$

Where x is state vector; u is the input; y is the output; f_n and g are smooth vector fields; h is smooth scalar function.

In order to obtain the input-output linearization of the multi-input multi-output (MIMO) system, the output y of the system is differentiated until the inputs appear.

$$\dot{y} = L_f \times h(x) + L_g \times h(x) \times u \tag{37}$$

where: $L_f \times h(x) = \dfrac{\partial h}{\partial x} \times f(x) \times L_g \times h(x) = \dfrac{\partial h}{\partial x} \times g(x);$ represent Lie derivatives of h(x) with respect to f(x) and g(x) respectively.

If $L_g \times h(x) = 0$ then the input u don't appear and the output is differentiated respectively.

$$y^{(r)} = L_f^r \times h(x) + L_g^{r-1} \times h(x) \times u \tag{38}$$

By neglecting resistances of the stator phases the stator voltage will be expressed by:

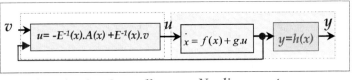

Figure 15. Schematic diagram of input-output linearizing control.

Where r is the relative rank of y. if we perform the above procedure for each input y_i, we get a total of m equations in the above form, which can be written completely as:

$$\begin{bmatrix} y_1^{(r)} \\ \dots \\ \dots \\ y_m^{(rm)} \end{bmatrix} = A(x) + E(x). \begin{bmatrix} u_1 \\ \dots \\ \dots \\ u_m \end{bmatrix} \quad (39)$$

Where the m*m matrix $E(x)$ is defined as:

$$E(x) = \begin{bmatrix} L_{g1} L_f^{r-1}.h_1 & \dots & \dots & L_{gm} L_f^{r-1}.h_1 \\ \dots & \dots & \dots & \dots \\ \dots & \dots & \dots & \dots \\ L_{g1} L_f^{rm-1}.h_m & \dots & \dots & L_{gm} L_f^{rm-1}.h_m \end{bmatrix} \quad (40)$$

$$A(x) = \begin{bmatrix} L_f^r.h_1 & \dots & \dots & L_f^{rm}.h_{1m} \end{bmatrix}^T \quad (41)$$

The matrix $E(x)$ is the decoupling matrix for the system. If $E(x)$ is nonsingular, then the original input u is controlled by the coordinate transformation:

$$u = -E^{-1}(x).A(x) + E^{-1}(x).v \quad (42)$$

Where: $v = \begin{bmatrix} v_1 & \dots & \dots & v_m \end{bmatrix}^T$. Substituting (41) into (39) obtains a linear differential relation between the output y and the new input v.

$$
\begin{bmatrix} y_1^{(r)} \\ \cdots \\ \cdots \\ y_m^{(rm)} \end{bmatrix} = \begin{bmatrix} v_1 \\ \cdots \\ \cdots \\ v_m \end{bmatrix} \tag{43}
$$

6.2. Control of Rotor Side Converter

According (33) and (34), the direct and quadrature components of the stator and the rotor currents are linearly dependent respectively, thus we chooses state vectors of the DFIG as follows

$$
x = \begin{bmatrix} x_1 & x_1 \end{bmatrix}^T = \begin{bmatrix} i_{rd} & i_{rq} \end{bmatrix}^T \tag{44}
$$

By substituting (24), (26), (27), (28) and (30) to (22) and (23), the following equations hold:

$$
V_{rd} = R_r.i_{rd} + \sigma.L_r.\frac{di_{rd}}{dt} - (\omega_s - \omega_r).\sigma.i_{rq} \tag{45}
$$

$$
V_{rq} = R_r.i_{rq} + \sigma.\frac{di_{rq}}{dt} - (\omega_s - \omega_r).\sigma.L_r.i_{rd} \tag{46}
$$

where: $\sigma = 1 - \dfrac{L_m^2}{L_s.L_r}$

Arranging (45) and (46) in the form of (32)

$$
\frac{di_{rd}}{dt} = -\frac{R_r}{\sigma.L_r}.i_{rd} + \frac{1}{L_r}.(\omega_s - \omega_r).i_{rq} + \frac{V_{rd}}{\sigma.L_r} \tag{47}
$$

$$
\frac{di_{rq}}{dt} = -\frac{R_r}{\sigma}.i_{rq} - (\omega_s - \omega_r).L_r.i_{rd} + \frac{V_{rq}}{\sigma} \tag{48}
$$

Defining the input of the DFIG system:

$$u = [u_1 \quad u_1]^T = [V_{rd} \quad V_{rq}]^T \tag{49}$$

From (44) and (45), we have:

$$f_{n1} = -\frac{R_r}{\sigma.L_r}.i_{rd} + \frac{1}{L_r}.(\omega_s - \omega_r).i_{rq} \tag{50}$$

$$f_{n2} = -\frac{R_r}{\sigma}.i_{rq} - L_r.(\omega_s - \omega_r).i_{rd} \tag{51}$$

$$g = \begin{bmatrix} \dfrac{1}{\sigma.L_r} & 0 \\ 0 & \dfrac{1}{\sigma} \end{bmatrix} \tag{52}$$

Since the rotor side controller is set to decouple the active and reactive powers, the active and reactive powers of stator are selected as the output:

$$g = \begin{bmatrix} \dfrac{1}{\sigma.L_r} & 0 \\ 0 & \dfrac{1}{\sigma} \end{bmatrix} \tag{53}$$

$$y = \begin{bmatrix} y_1 \\ y_2 \end{bmatrix} = \begin{bmatrix} P_s \\ Q_s \end{bmatrix} = \begin{bmatrix} V_{sd}.i_{sd} + V_{sq}.i_{sq} \\ V_{sq}.i_{sd} - V_{sd}.i_{sd} \end{bmatrix} \tag{54}$$

From (33), (34) and (54):

$$y_1 = \frac{\phi_s}{L_s}.V_{sd} - \frac{L_m}{L_s}.(V_{sd}.i_{rd} - V_{sq}.i_{rq}) \tag{55}$$

$$y_2 = \frac{\phi_s}{L_s}.V_{sq} - \frac{L_m}{L_s}.(V_{sq}.i_{rd} - V_{sd}.i_{rq}) \tag{56}$$

Differentiating (55) and (56) until an input appears:

$$\dot{y}_1 = \frac{\dot{V}_{sd}}{L_s}.(\phi_s - L_m.i_{rd}) - \frac{L_m}{L_s}.\dot{V}_{sq}.i_{rq} - \frac{L_m}{L_s}.(V_{sd}.f_{n1} - V_{sq}.f_{n2}) - \frac{L_m.V_{sd}}{\sigma.L_r.L_s}.V_{rd} - \frac{L_m.V_{sq}}{\sigma.L_s}.V_{rq}$$

(57)

$$\dot{y}_2 = \frac{\dot{V}_{sq}}{L_s}.(\phi_s - L_m.i_{rd}) - \frac{L_m}{L_s}.\dot{V}_{sd}.i_{rd} - \frac{L_m}{L_s}.(V_{sq}.f_{n1} - V_{sq}.f_{n2}) - \frac{L_m.V_{sq}}{\sigma.L_r.L_s}.V_{rd} + \frac{L_m.V_{sd}}{\sigma.L_s}.V_{rq}$$

(58)

Rewriting (54) and (58) in the form of (39)

$$\begin{bmatrix} \dot{y}_1 \\ \dot{y}_2 \end{bmatrix} = A(x) + E(x).\begin{bmatrix} u_1 \\ u_2 \end{bmatrix}$$

(59)

Where:

$$A(x) = \begin{bmatrix} \frac{\dot{V}_{sd}}{L_s}.(\phi_s - L_m.x_1) - \frac{L_m}{L_s}.\dot{V}_{sq}.x_2 - \frac{L_m}{L_s}.(V_{sd}.f_{n1} + V_{sq}.f_{n2}) \\ \frac{\dot{V}_{sq}}{L_s}.(\phi_s - L_m.x_1) + \frac{L_m}{L_s}.\dot{V}_{sd}.x_2 - \frac{L_m}{L_s}.(V_{sq}.f_{n1} - V_{sq}.f_{n2}) \end{bmatrix}$$

(60)

$$E(x) = \begin{bmatrix} -\dfrac{L_m.V_{sd}}{\sigma.L_r.L_s} & -\dfrac{L_m.V_{sq}}{\sigma.L_s} \\ -\dfrac{L_m.V_{sq}}{\sigma.L_r.L_s} & \dfrac{L_m.V_{sd}}{\sigma.L_s} \end{bmatrix}$$

(61)

Since $E(x)$ is nonsingular, the control scheme is given from (58) and (59) as:

$$\begin{bmatrix} V_{rd} \\ V_{rq} \end{bmatrix} = E^{-1}(x).\begin{bmatrix} -A(x) + \begin{pmatrix} u_1 \\ u_2 \end{pmatrix} \end{bmatrix}$$

(62)

In $A(x)$ and $E(x)$ most components relate to the factors Lm/Ls and σ that both are equal to one approximately, and the functions f_{n1} and f_{n2} have no relation with parameters of the stator windings. Hence, the control law is robust to machine parameter variations.

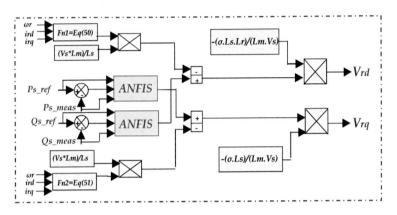

Figure 16. Schematic diagram of input/output linearizing control of DFIG.

For a DFIG the demanded active power Ps^{*4} is decided by the maximum power tracking scheme of the wind turbine according to the instant wind speed, and the demanded reactive power Qs^* is set by the grid operator to support grid voltage. The NFC achieves the tracking of the powers. The proposed control is described in details in Figure 16.

7. GRID SIDE CONVERTER (GSC) AND DC-LINK VOLTAGE CONTROL

Figure 17 presents the *GSC* configuration. In this case, we use balanced network voltages, so we will have the following relationships (Youcef Bekakra 2014; Julius Mwaniki 2017):

---

[4] In this chapter; Ps* = Ps_ref (reference stator active power) & Qs* = Qs_ref (reference stator reactive power).

$$\begin{cases} V_{1N} = \dfrac{1}{3}.(+2.V_1 - V_2 - V_3) \\ V_{2N} = \dfrac{1}{3}.(-V_1 + 2V_2 - V_3) \\ V_{3N} = \dfrac{1}{3}.(-V_1 - V_2 + 2V_3) \end{cases} \tag{63}$$

According to the closing or the opening of the switches[5] T_{ij}, the voltages of branch (*leg*) V_i can be equal to V_{dc} or 0. Other variables such as S_{11}, S_{12} and S_{13} are introduced which take 1 if the switch T_{ij} is closed or 0 if it is blocked. Equation (63) can be rewritten as:

$$\begin{pmatrix} V_{1N} \\ V_{2N} \\ V_{3N} \end{pmatrix} = \dfrac{V_{dc}}{3} \begin{pmatrix} +2 & -1 & -1 \\ -1 & +2 & -1 \\ -1 & -1 & +2 \end{pmatrix} \begin{pmatrix} S_{11} \\ S_{12} \\ S_{13} \end{pmatrix} \tag{64}$$

The rectified current can be written as:

$$I_{rec} = S_{11}.I_{ga} + S_{12}.I_{gb} + S_{13}.I_{gc} \tag{65}$$

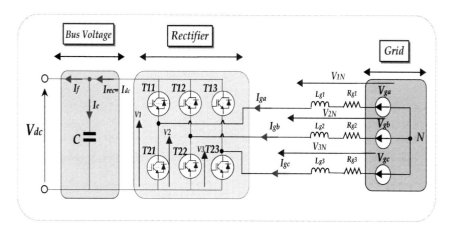

Figure 17. GSC configuration.

---

[5] Knowing that: T_{ij} (i = 1, 2 'lines' and j = 1, 2, 3 'columns').

Where S_{li} presents a logical signal deduced from the application of the control technique of *PWM*. In this section, the switching signals are determined by the comparison (*using hysteresis controllers*) between the measured grid currents I_{g_abc} and the reference grid currents $I^*_{g\_abc}$.

The terminal voltage of the capacitor is calculated by:

$$C.\frac{dV_{dc}}{dt} = I_C = I_{rec} = I_f = (S_{11}.I_{ga} + S_{12}.I_{gb} + S_{13}.I_{gc}) - I_f \quad (66)$$

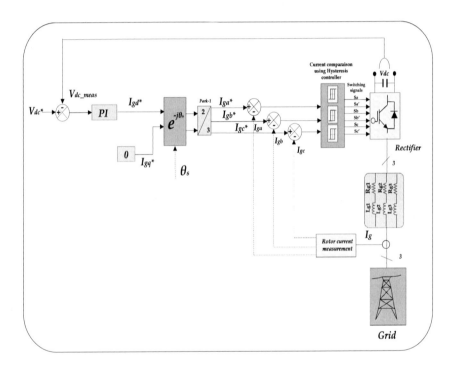

Figure 18. Grid side converter topology (DC-link voltage control).

Figure 18 displays the control block diagram of a vector control strategy for the *GSC*. This grid *PWM converter* is operated to keep the *DC-link* voltage at a constant value. The *GSC* is usually controlled with a vector control strategy with the grid voltage orientation. This voltage frame corresponds to the *d-q axes*, which makes it possible to decouple the expressions from the active and the reactive power exchanged between the *grid* and the *rotor side*. A *DC* capacitor is used in order to remove ripple and keep the *DC-link* voltage relatively smooth. Therefore, a hysteresis controller is used in which the error between the desired and actual currents is passed through a controller (Aman Abdulla Tanvir 2015).

The control of active power and consequent control of the *DC-link* voltage are realized by the intermediary of reference direct grid current I^*_d and the reactive power by the intermediary of reference transversal grid current I^*_{gq}. In order to guarantee the unity power factor ($PF \approx 1$) at the *grid- side*; the reference transversal grid current I^*_{gq} is maintained to zero value ($I^*_{gq} = 0\ (A)$).

8. ROTOR SIDE CONVERTER (RSC)

The control of the rotor side converter (*RSC*) enables us to control the stator active and reactive powers independently. From equation (35), it's clear that the active and reactive powers are based on the *q* and *d axes* rotor currents respectively.

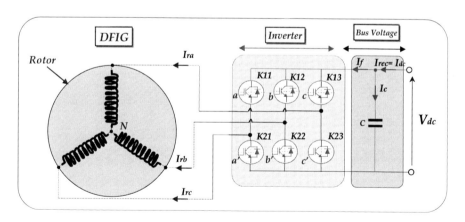

Figure 19. RSC configuration (three-phase voltage source PWM Inverter+ DFIG).

Therefore, the powers are checked by controlling the rotor currents. These currents are controlled by *PI* controllers. Then, we must add terms of compensation and decoupling (Figure 19 *illustrates the global RSC configuration*). The voltages obtained are transformed to '*abc*' frame using '*dq*' to '*abc*' transformation which the angle is the difference between the stator angle obtained using *PLL (Phase locked Loop)*[6] and the rotor angle (*as shown in* Figure 20). Finally, the '*abc*' results voltages are transformed to '*Alpha-Beta*' and by the means of the *DC* bus voltage *and* sectors; it can be converted to *SVM (Space vector modulation)* signals in order to control the gates of *IGBTs* used in the rotor side converter (*RSC*) as set forth in Figure 19 (M. Itsaso Martinez 2012; A. Susperregui 2010).

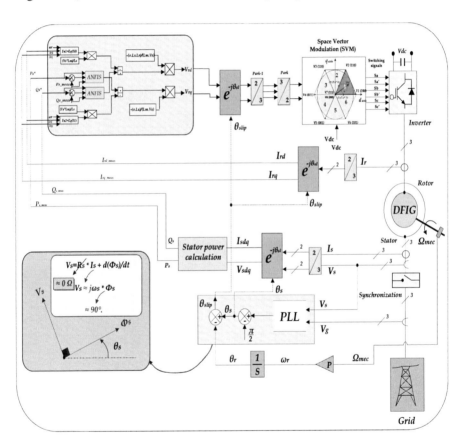

Figure 20. Rotor side converter topology (of Indirect Power Control DFIG).

---

[6] Please refer to chapter.6 for more information (paragraph § 6.6.2; a robust PLL is described with descriptive schemes).

8.1. Space Vector Modulation (SVM)

The circuit model of a typical three-phase voltage source *PWM* (*Pulse width modulation*)[7] inverter is shown in Figure 19. K_{11} to K_{23} are the six power switches[8] that shape the output, which are controlled by the switching variables *a, a', b, b', c and c'*. When an upper transistor is switched on, i.e., when *a, b* or *c* is *1*, the corresponding lower transistor is switched off, i.e., the corresponding *a', b' or c'* is *0*. Therefore, the ON and OFF states of the upper transistors K_{11}, K_{12} and K_{13} can be used to determine the output voltage (Keliang Zhou 2002; Behzad Vafakhah 2010).

The relationship between the switching variable vector $[a, b, c]^t$ and the line-to-line voltage vector $[V_{rab}\ V_{rbc}\ V_{rca}]^t$ is given by equation (67) in the following:

$$\begin{pmatrix} V_{rab} \\ V_{rbc} \\ V_{rca} \end{pmatrix} = V_{dc} \cdot \begin{pmatrix} +1 & -1 & 0 \\ 0 & +1 & -1 \\ -1 & 0 & +1 \end{pmatrix} \begin{pmatrix} a \\ b \\ c \end{pmatrix} \tag{67}$$

Also, the relationship between the switching variable vector $[a, b, c]^t$ and the phase voltage vector $[V_a\ V_b\ V_c]^t$ can be expressed below (equation (68)).

$$\begin{pmatrix} V_{rab} \\ V_{rbc} \\ V_{rca} \end{pmatrix} = \frac{V_{dc}}{3} \cdot \begin{pmatrix} +2 & -1 & -1 \\ -1 & +2 & -1 \\ -1 & -1 & +2 \end{pmatrix} \begin{pmatrix} a \\ b \\ c \end{pmatrix} \tag{68}$$

As illustrated in Figure 21, there are eight possible combinations of on and off patterns for the three upper power switches. The on and off states of the lower power devices are opposite to the upper one and so are easily determined once the states of the upper power transistors are determined. According to equations (67) & (68), the eight switching vectors, output

---

[8] Knowing that: K_{ij} (i = 1, 2 'are the lines' and j = 1, 2, 3 'are the columns').

line to neutral voltage (*phase voltage*), and output line-to-line voltages in terms of *DC-link* V_{dc}, are given in Table 2 and Figure 21 shows the eight inverter voltage vectors (V_0 to V_7).

Space Vector *PWM* (*SVPWM*) refers to a special switching sequence of the upper three power transistors of a three-phase power inverter. It has been shown to generate less harmonic distortion in the output voltages and or currents applied to the phases of an *AC* motor and to provide more efficient use of supply voltage compared with sinusoidal modulation technique.

As described in Figure 22, this transformation is equivalent to an orthogonal projection of *[a, b, c]^t* onto the two-dimensional perpendicular to the vector *[1, 1, 1]^t* (*the equivalent d-q plane*) in a three-dimensional coordinate system. As a result, six non-zero vectors and two zero vectors are possible. Six non-zero vectors (V_1 - V_6) shape the axes of a hexagonal as depicted in Figure 23, and feed electric power to the load.

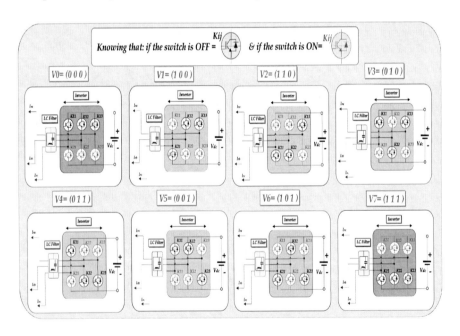

Figure 21. The eight inverter voltage vectors (V_0 to V_7).

Table 2. Switching vectors, phase voltages and output line to line voltages

Voltage vectors	Switching vectors: a	b	c	Line to neutral voltage: V_{an}	V_{bn}	V_{cn}	Line to line voltage: V_{ab}	V_{bc}	V_{ca}
V_0	0	0	0	0	0	0	0	0	0
V_1	1	0	0	2/3	-1/3	-1/3	+1	0	-1
V_2	1	1	0	1/3	1/3	-2/3	0	+1	-1
V_3	0	1	0	-1/3	2/3	-1/3	-1	+1	0
V_4	0	1	1	-2/3	1/3	1/3	-1	0	+1
V_5	0	0	1	-1/3	-1/3	2/3	0	-1	+1
V_6	1	0	1	1/3	-2/3	1/3	+1	-1	0
V_7	1	1	1	0	0	0	0	0	0

NB: The respective voltage should be multiplied by V_{dc}.

The angle between any adjacent two non-zero vectors is *sixty* degrees (*60°*). Mean-while, two zero vectors (V_0 and V_7) are at the origin and apply zero voltage to the load. The eight vectors are called the basic space vectors and are denoted by V_0, V_1, V_2, V_3, V_4, V_5, V_6, and V_7.

The same transformation can be applied to the desired output voltage to get the desired reference voltage vector V_{ref} in the *d-q* plane. The objective of space vector *PWM* technique is to approximate the reference voltage vector V_{ref} using the eight switching patterns. One simple simple method of approximation is to generate the average output of the inverter in a small period, *T* to be the same as that of V_{ref} in the same period. Knowing that, the main aim of using SVM strategy is to fix the switching frequency *IGBTs*.

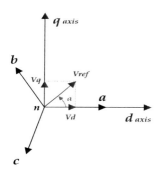

Figure 22. Voltage space vector and its components in d & q axes.

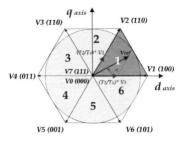

Figure 23. The eight (08) basic switching vectors and sectors.

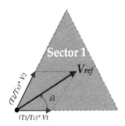

Figure 24. Reference vector as a combi-nation of adjacent vectors at sector 1.

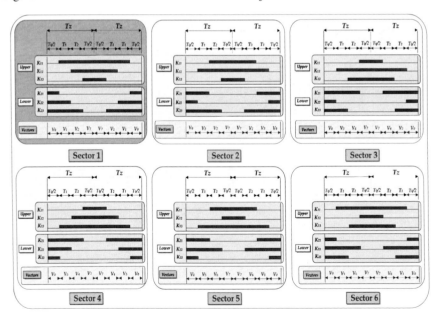

Figure 25. Space vector PWM switching patterns at each sector.

Novel I/O-LDC Control Based on ANFIS ...   79

Figure 25 shows space vector *PWM* switching patterns at each sector. The switching time at each sector is summarized in Table 3, and it will be built in *Simulink model* to implement *SVM*.

Table 3. Switching time calculation at each sector

Sector	Upper switches (K_{11}, K_{12}, K_{13})	Lower switches (K_{21}, K_{22}, K_{23})
1	$K_{11} = T_1 + T_2 + T_0/2$	$K_{21} = T_0/2$
	$K_{12} = T_2 + T_0/2$	$K_{22} = T_1 + T_0/2$
	$K_{13} = T_0/2$	$K_{23} = T_1 + T_2 + T_0/2$
2	$K_{11} = T_1 + T_0/2$	$K_{21} = T_2 + T_0/2$
	$K_{12} = T_1 + T_2 + T_0/2$	$K_{22} = T_0/2$
	$K_{13} = T_0/2$	$K_{23} = T_1 + T_2 + T_0/2$
3	$K_{11} = T_0/2$	$K_{21} = T_1 + T_2 + T_0/2$
	$K_{12} = T_1 + T_2 + T_0/2$	$K_{22} = T_0/2$
	$K_{13} = T_2 + T_0/2$	$K_{23} = T_1 + T_0/2$
4	$K_{11} = T_0/2$	$K_{21} = T_1 + T_2 + T_0/2$
	$K_{12} = T_1 + T_0/2$	$K_{22} = T_2 + T_0/2$
	$K_{13} = T_1 + T_2 + T_0/2$	$K_{23} = T_0/2$
5	$K_{11} = T_2 + T_0/2$	$K_{21} = T_1 + T_0/2$
	$K_{12} = T_0/2$	$K_{22} = T_1 + T_2 + T_0/2$
	$K_{13} = T_1 + T_2 + T_0/2$	$K_{23} = T_0/2$
6	$K_{11} = T_1 + T_2 + T_0/2$	$K_{21} = T_0/2$
	$K_{12} = T_0/2$	$K_{22} = T_1 + T_2 + T_0/2$
	$K_{13} = T_1 + T_0/2$	$K_{23} = T_2 + T_0/2$

Therefore, space vector *PWM* can be implemented by the following steps:

- **Step 1:** *Determine V_d, V_q, V_{ref}, and angle (α):*

From Figure 22, the V_d, V_q, V_{ref}, and angle (*α*) can be determined as follows:

$$V_d = V_{an} - V_{bn}.\cos(60°) - V_{cn}.\cos(60°) = V_{an} - \frac{1}{2}.V_{bn} - \frac{1}{2}.V_{cn} \qquad (69)$$

$$V_q = 0 - V_{bn}.\cos(30°) - V_{cn}.\cos(30°) = V_{an} + \frac{\sqrt{3}}{2}.V_{bn} - \frac{\sqrt{3}}{2}.V_{cn} \qquad (70)$$

$$\begin{pmatrix} V_d \\ V_q \end{pmatrix} = \frac{2}{3} \cdot \begin{pmatrix} 1 & -\dfrac{1}{2} & -\dfrac{1}{2} \\ 0 & +\dfrac{\sqrt{3}}{2} & -\dfrac{\sqrt{3}}{2} \end{pmatrix} \begin{pmatrix} V_{an} \\ V_{bn} \\ V_{cn} \end{pmatrix} \tag{71}$$

$$\left\| \overrightarrow{V_{ref}} \right\| = \sqrt{V_d^2 + V_q^2} \tag{72}$$

$$a = \tan^{-1}\left(\frac{V_q}{V_d}\right) = \omega.t = 2.\pi.f \tag{73}$$

Where 'f' is fundamental frequency.

- **Step 2:** *Determine time duration T_1, T_2, T_0:*
  From Figure 24, the switching time duration can be calculated as follows:
  Switching time duration at Sector 1:

$$\int_0^{T_z} \overrightarrow{V_{ref}} = \int_0^{T1} \overrightarrow{V_1} dt + \int_{T1}^{T1+T2} \overrightarrow{V_2} dt + \int_{T1+T2}^{TZ} \overrightarrow{V_0} dt \tag{74}$$

$$T_Z \overrightarrow{V_{ref}} = (T_1.\overrightarrow{V_1} + T_2.\overrightarrow{V_2}) \tag{75}$$

$$T_Z \left\| \overrightarrow{V_{ref}} \right\| \begin{pmatrix} \cos(a) \\ \sin(a) \end{pmatrix} = T_1.\frac{2}{3}.V_{dc}\begin{pmatrix} 1 \\ 2 \end{pmatrix} + T_2.\frac{2}{3}.V_{dc}\begin{pmatrix} \cos(\pi/3) \\ \sin(\pi/3) \end{pmatrix} \tag{76}$$

Where: $0° \leq a \leq 60°$.

$$T_1 = T_Z.a.\frac{\sin(\pi/3 - a)}{\sin(\pi/3)} \tag{77}$$

$$T_2 = T_Z.a.\frac{\sin(a)}{\sin(\pi/3)} \tag{78}$$

$$T_0 = T_Z - (T_1 + T_2) \tag{79}$$

Where: $T_z = (1/f_z)$ and $a = \left\| \overrightarrow{V_{ref}} \right\| / \left(\frac{2}{3} . V_{dc} \right)$

Switching time duration at any Sector:

$$T_1 = \frac{\sqrt{3}.T_Z.\left\| \overrightarrow{V_{ref}} \right\|}{V_{dc}}.(\sin(\frac{\pi}{3} - a + \frac{n-1}{3}.\pi)) = \frac{\sqrt{3}.T_Z.\left\| \overrightarrow{V_{ref}} \right\|}{V_{dc}}.(\sin(\frac{n}{3}.\pi - a)) = \frac{\sqrt{3}.T_Z.\left\| \overrightarrow{V_{ref}} \right\|}{V_{dc}}.$$
$$(\sin(\frac{n}{3}).\pi.\cos(a) - \cos(\frac{n}{3}).\pi.\sin(a) \tag{80}$$

$$T_2 = \frac{\sqrt{3}.T_Z.\left\| \overrightarrow{V_{ref}} \right\|}{V_{dc}}.(\sin(a - \frac{n-1}{3}.\pi)) = \frac{\sqrt{3}.T_Z.\left\| \overrightarrow{V_{ref}} \right\|}{V_{dc}}.(-\cos(a).\sin(\frac{n-1}{3}).\pi + \sin(a).\cos(\frac{n-1}{3}).\pi) \tag{81}$$

$$T_0 = T_z - (T_1 + T_2) \tag{82}$$

Where, n = 1 through 6 (that is: Sector 1 to 6) and $0° \leq a \leq 60°$.

- **Step 3:** *Determine the switching time of each transistor (K_{11} to K_{23})*

9. Operating Principle of DFIG

In spite of the disadvantages associated with the slip-rings, the wound-rotor induction machine has long been a wind electric generator choice. By using a suitable integrated approach in the design of a *WECS*, use of a slip-ring induction generator has been found to be economically competitive. Control of grid-connected and isolated variable-speed wind turbines with a doubly fed induction generator has been implemented and reported (Gonzalo Abad Biain 2008; Gonzalo Abad 2014).

A wound-rotor induction machine can be operated as a doubly-fed induction machine (*DFIM*) when a power converter is present in its rotor circuit. This converter directs the power flow into and out of the rotor windings. The *DFIM* could operate as either a motor or a generator at sub-synchronous and super-synchronous speeds[9], there exist four (04) operational modes. All the four modes are explained in Figure 26. When the machine runs above synchronous speed, this operation is termed super-synchronous operation. Similarly, operation below synchronous speed is called sub-synchronous operation. In both sub- & super-synchronous operation, the machine can be operated either as a motor or a generator. In the motoring mode of operation, the torque produced by the machine is positive.

On the other hand, during generating operation, the machine needs mechanical torque as input; thus, the torque is negative during generating operation. The principle of a DFIM control in these modes can be understood more clearly by the power-flow diagram given in Figure 28. In this figure, P_s is the stator power, P_r is the rotor power, and P_m is the mechanical power. When the *DFIM* is operating as a motor in the sub-synchronous speed range (Figure 28-[1]), power is taken out of the rotor. This operational mode is commonly known as slip-power recovery. If the speed increases so that the machine is operating at super- synchronous speeds (Figure 28-[2]), the rotor power then changes direction (Gonzalo Abad Biain 2008; Gonzalo Abad 2014]. When the *DFIM* is operating as a generator in the sub-synchronous speed range (Figure 28[3]), power is delivered to the rotor. If the speed increases so that the machine is operating at super-synchronous speeds (Figure 28-[4]) the rotor power again changes direction.

Because the machine will be predominantly working as a generator for a wind-energy application, operation in *Mode 3* and *Mode 4* is more important than *Mode 1* and *Mode 2*. However, for an ideal *WECS* system,

---

[9] In French literature, Sub-synchronous and Super-synchronous speeds are replaced by "Hypo-synchrone and Hyper-synchrone".

all four operating modes are desirable. Motoring modes are useful when the generator needs to speed up quickly in order to achieve the best operating speed and efficiency (Gonzalo Abad Biain 2008; Gonzalo Abad 2014). However, because of the high inertia of the wind generator power train, the acceleration of the machine may be achieved by the wind torque itself. Hence, motoring operations may be sacrificed if the cost of the system can be reduced substantially. Figure 27 displays the *Torque/Speed* characteristic of *DFIM* in four (04) quadrants under both modes: *Motoring and Generating*. It should be noted that for Sub- and Super-synchronous generating modes, the power flows through the rotor are of opposite directions. Hence, the power converter connected with such system should have bi-directional power flow capability.

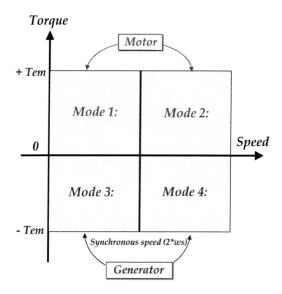

Figure 26. Operating modes of a DFIM: Mode 1 & 2 (Sub and Super synchronous motoring mode), Mode 3 & 4 (Sub and Super synchronous generating mode).

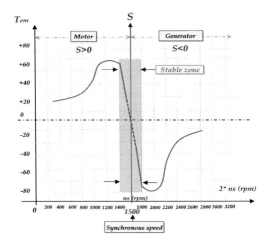

Figure 27. DFIM's torque/speed characteristic.

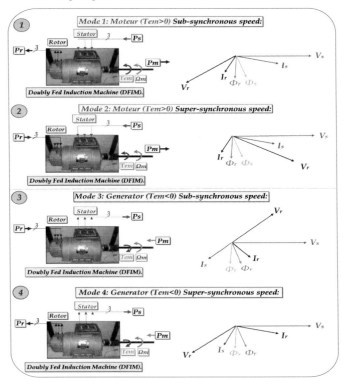

Figure 28. Power-flow diagram of a DFIM for (1): Sub-synchronous motoring mode, (2): Super-synchronous motoring mode, (3): Sub-synchronous generating mode, and (4): Super-synchronous generating mode.

10. SIMULATION RESULTS

The proposed system (DFIG control + wind turbine)[10] is validated using Matlab R2009a/Simulink® software under MPPT strategy (Mode-2) by keeping stator reactive power equals to zero and to ensure unity power factor (PF = 1).

The figures-(29 to 33) respectively present the simulation results of proposed I/OLDC using PI & NFC-knowing that the wind-system based on DFIG (4 kW) and wind turbine (4.5 kW)-. These figures are divided into two parts;

- The first part depicts the behavior of the wind-system parameters under two (02) modes[11] as follows:
  1. *Mode 1 (Red color*/Figure 29): Without MPPT Strategy, in this case we impose the P_s and Q_s reference profiles (as shown in Table 4).
  2. *Mode 2 (Green color*/Figure 30): With MPPT strategy, in this case we propose a medium wind speed based on random form (*Max wind speed = 13.5 m/sec*) by keeping stator reactive power equal to Zero level "$Q_s \approx 0$ Var"; to ensure only the exchange of the stator active power to the grid; means to follow (track) the maximum active power point despite the sudden wind speed variation.

Table 4. The proposed profiles of the stator active and reactive power references

Time (sec):	Stator active power (W):	Stator reactive power (Var):
[0 - 0.3]	-700.	0.
[0.3–0.7]	-2000.	+1000.
[0.9–1.0]	-2800.	0.
[1.0–1.2]	-700.	-1400.
[1.2–1.5]	-700.	0.

---

[10] Please refer to Table.6 and Table.7 respectively).
[11] Knowing that for each Mode we present the conventional and novel I/OLDC.

86 Fayssal Amrane and Azeddine Chaiba

- The second part deals on robustness tests using a comparative simulation study, for both modes (with/without MPPT strategy) are described in details in figures.32 and Figure 33 respectively. This section is developed in order to verify the robustness of wind-system under parameter variation (*using three tests*)[12] in transient *and* steady states.

10.1. Mode 1 (Proposed I/OLDC Using PI & NFC, without MPPT Strategy)

Figure 29 shows the behavior of wind-system parameters of *Mode 1* under two topologies; the measured stator active and reactive powers (P_{s_meas} and Q_{s_meas}) and their references (P_{s_ref} and Q_{s_ref}) profiles are presented together in Figure 29-(a) and are presented separately in figures.29.(b and c) respectively. The reference powers are indicated in Table 4. The direct and quadrature components of currents and flux (I_{rd}, I_{rq} and Φ_{rd}, Φ_{rq}) are presented respectively in Figure 29-(e and g), which present the inverse diagrams compared to reactive and active powers. The inverse case for stator direct and quadrature currents (I_{sd}, I_{sq}) which have the same diagrams of reactive and active powers, and they are presented in Figure 29-(d). The power error is presented in Figure 29-(f). The stator' and rotor currents; I_{s_abc} and I_{r_abc} are shown in Figure 29-(h and i) respectively, we remark the sinusoidal form of the three rotor and stator phases currents, have an excellent THD of stator currents (<< *5% respect the IEEE-519 Std*) for NFC topology compared to PI. The power factor (PF) of the conventional and proposed control are presented in Figure 29-(j), knowing that PF is the ratio of P to S (apparent power), had the top value when the reactive power equals to zero value (Qs = 0 var); refer to these time interval in figures.29-(a, b, c and j): 0-0.3 (sec), 0.9-1.0 (sec),

---

[12] Knowing that in this chapter the robustness tests are based on three tests as follows: [*Test-1*: without parameter changement → Blue color, *Test-2*: +100% of R_r and -25% of (L_s, L_r and L_m) → Brown color and *Test-3*: +100% of (J and R_r), -25 % of (L_r, L_s and L_m) → Green color] respectively.

Novel I/O-LDC Control Based on ANFIS ...

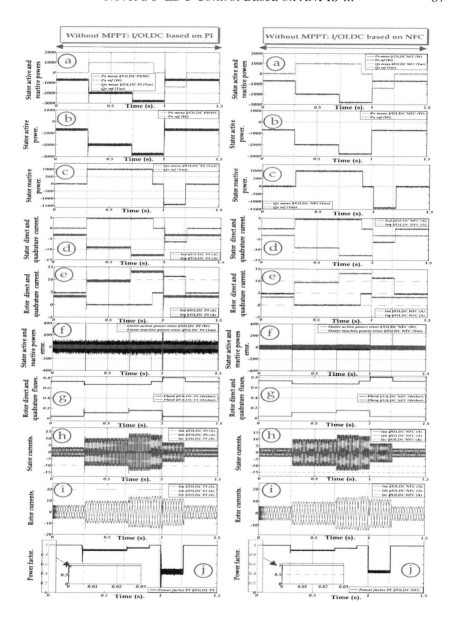

Figure 29. Simulation results of Mode-1(without MPPT) using PI & NFC respectively; (a): stator active and reactive powers, (b): stator active power, (c): stator reactive power, (d): stator direct and quadrature currents, (e): rotor direct and quadrature currents, (f): stator active and reactive power error, (g): rotor direct and quadrature fluxes, (h): stator currents, (i): rotor currents, (j): power factor.

0.8-1.0 and 1.2-1.5(sec). It is noted that the measured powers have good tracking power (*refer to Figure 29-(b and c) to the right side*), a few ($\approx$ 09%) overshoot is noted is noted in interval: 0.3 and 1.0 (sec) respectively in P_{s_meas} and Q_{s_meas}. A good THD of stator currents will be injected into the grid (*= 01.19% respect the IEEE-519 Std*) and a small power errors are noted: -120 (W_Var) $\leq \Delta P_s \Delta Q_s \leq$ +120 (W_Var). The obtained results of novel and conventional I/OLDC based on NFC and PI under mode-1; are shown in Table 5.

Table 5. Performances results for proposed control using PI & NFC respectively

		THD_Is_abc (%):	THD_Ir_abc (%):	Overshoot:	Response time (sec):	Power Error (W_Var):
PI	*Mode 1:*	0.76%	177.25%	Remarkable ($\approx$ 20%).	2.3 * 10⁻⁴.	+/- 150.
	Mode 2:	0.31%	05.41%	Few ($\approx$ 10%).	2.5* 10⁻⁴.	+/- 150.
NFC	*Mode 1:*	01.19%	88.20%	Few ($\approx$ 9%).	0.33 * 10-3.	+/- 60.
	Mode 2:	0.56%	05.26%	Neglected (<5%).	0.37 * 10-3.	+/- 60.

10.2. Mode 2 (Proposed I/OLDC Using PI & NFC, with MPPT Strategy)

Figure 30 shows the behavior of wind-system parameters of *Mode 2* under tow topologies; the reference stator active power (P_{s_ref}) (Figure 30-(a)) is extracted from MPPT strategy (in this case, the wind speed will take random form); it takes the inverse diagram of wind speed. The stator reactive power (Q_{s_ref}) equals to 0 (Var), represents power factor unity. The measured stator active and reactive powers (P_{s_meas} and Q_{s_meas}) and their references (P_{s_ref} and Q_{s_ref}) profiles are presented together in Figure 30-(a) and are presented separately in figures.30-(b and c) respectively. The direct and quadrature components of currents and flux (I_{rd}, I_{rq} and Φ_{rd}, Φ_{rq}) are presented respectively in Figure 30-(e and g), which present the inverse

Novel I/O-LDC Control Based on ANFIS ...

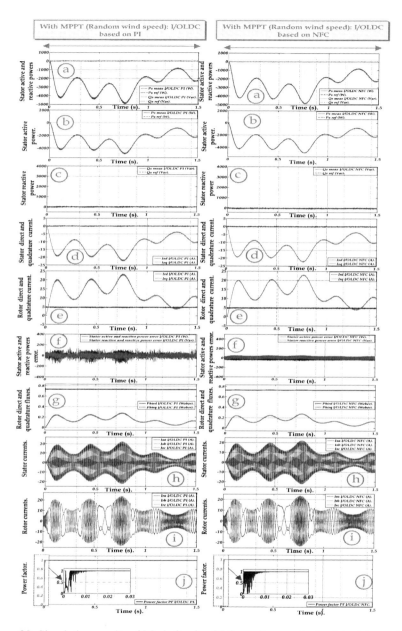

Figure 30. Simulation results of Mode 2 using PI & NFC respectively; (a): stator active and reactive powers, (b): stator active power, (c): stator reactive power, (d): stator direct and quadrature currents, (e): rotor direct and quadrature currents, (f): stator active and reactive power error, (g): rotor direct and quadrature fluxes, (h): stator currents, (i): rotor currents, (j): power factor.

90   Fayssal Amrane and Azeddine Chaiba

diagrams compared to reactive and active powers. The inverse case for stator direct and quadrature currents (I_{sd}, I_{sq}) which have the same diagrams of reactive and active powers, and they are presented respectively in Figure 30-(d). The power error is presented in Figure 30-(f). The stator' and rotor currents; I_{s_abc} and I_{r_abc} are shown in figures.30-(h and i) respectively, we remark the sinusoidal form of the waveforms and have an excellent THD of stator currents ($<<$ *5% respect the IEEE-519 Std*) for ANFIS topology (to the righ). The power factor (PF) of the conventional and proposed control is presented in Figure 30-(j), it reaches the top value when the reactive power equals to zero value, in this case the stator reactive power equals to zero value means the PF had taken always the unity.

An excellent tracking power is noted in transient and steady states, a neglected ($<5\%$) overshoot is noted at "0.2 (sec) to 0.3 (sec) and from 09 (sec) to 1.5 (sec)" in measured stator reactive' and active power (P_{s_meas} and Q_{s_meas}). An excellent THD of stator currents will be injected into the grid (= *0.31% for stator currents and 05.26% for rotor currents respecti- vely respect the IEEE-519 Std*). An acceptable power error is noted: -150 (W_Var) $\leq \Delta P_s\_\Delta Q_s \leq$ +150 (W_Var). Table 4 shows in details the obtained results of the conventional and novel I/OLDC based on PI and NFC under Mode-2.

10.3. Robustness Tests[13] for Mode 1 and Mode 2

Figures 32 (a,b)-33 (a,b) illustrate the behavior of measured active and reactive powers and theirs references respectively under parameters variations; in transient and steady states.

---

[13] Knowing that in this chapter the robustness tests are based on three tests as follows: [*Test-1*: without parameter changement $\rightarrow$ Blue color, *Test-2*: +100% of R_r and -25% of (L_s, L_r and L_m) $\rightarrow$ Brown color and *Test-3*: +100% of (J and R_r), -25 % of (L_r, L_s and L_m) $\rightarrow$ Green color] respectively.

Novel I/O-LDC Control Based on ANFIS ...   91

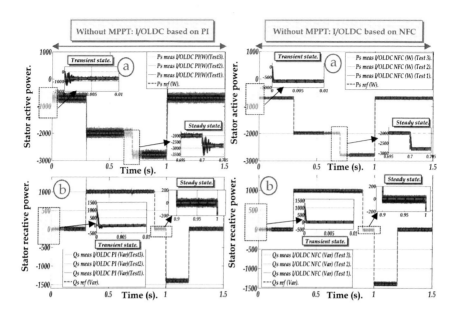

Figure 32. Robustness tests of proposed control under Mode 1 using PI and NFC respectively; (a): stator active powers, (b): stator reactive powers.

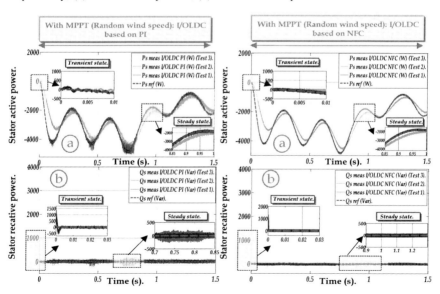

Figure 33. Robustness tests of proposed control under Mode 2 using PI and NFC respectively; (a): stator active powers, (b): stator reactive powers.

A. Mode 1 (Proposed I/OLDC Based on PI & NFC without MPPT Strategy)

It can be noted few power in active and reactive power (conventional I/OLDC based on PI/-please refer to the left side of Figure 32) especially using the 2nd test and 3rd test (green color) with small undulations espacially in transient and steady states (*please refer to zoom*) and the value of power error reaches nearly ± 120 (W_Var) in Test-1, and nearly ±200 (W_Var) in Test-2 & Test-3, and remarkable overhoot is noted.

For the second topology (based on NFC/refer to the right side of Figure 32) also a small power error is noted for the first Test (blu color), and the value of power error reaches nearly ± 150 (W_Var) in Test-1, and nearly ±175 (W_Var) in Test-2 and Test-3. A few overshoot is noted under all robustnees tests espacially at 0.2 (sec) and at 1.0 (sec).

B. Mode 2 (Proposed I/OLDC Based on PI & NFC with MPPT Strategy)

Figures.33-(conventional I/OLDC/refer to the left side) display the behavior of stator active and reactive powers under MPPT strategy by maintaining the reactive power equals to zero value. In this case the active power had taken the inverse random profile of wind speed. Using robustnees tests a big undulations are noted (using tests: 2 and 3) espacially at 0.75 (sec) and 0.8 (sec) which presents the over rated power of DFIG (P = 4 (kW) and the measured active power maintain 4.6 kW, also a high overshoot is noted in transient and steady states.

For the second topology (based on NFC/refer to right side of the Figure 33) the neglected ripples are noted (using tests: 2 & 3), an acceptable power error and overshoot (which means that the NFC controllers can maintain the unity power factor under parameters changing) are shown in stator reactive power especially in steady states (Please refer to the zoom).

CONCLUSION

In this chapter, a high performance of improved feedback linearization based on the model of DFIG has been proposed. According the nonlinear control theory, the control algorithm can be more easily realized in comparison with the vector control, because the design of the controller depends only on the nonlinear theory. ANFIS controllers are proposed to control stator active and recative powers (Ps & Qs) instead the conventional controllers (PI). The results obtained by the validation platform using the *MATLAB/Simulink R2009a*, demonstrate the feasibility of the proposed algorithm, which prove the improvement of propoed control (I/OLDC based on ANFIS) in termes of tracking power, power error, over-shoot & high response dynamic performances in transient and steady states.

Table 6. Parameters of the DFIG

Rated Power:	4 kWatts
Stator Resistance:	Rs = 1.2 Ω
Rotor Resistance:	Rr = 1.8 Ω
Stator Inductance:	Ls = 0.1554 H.
Rotor Inductance:	Lr = 0.1558 H.
Mutual Inductance:	Lm = 0.15 H.
Rated Voltage:	Vs = 220/380 V
Number of Pole pairs:	P= 2
Rated Speed:	N=1440 rpm
Friction Coefficient:	fDFIG=0.00 N.m/s
The moment of inertia	J = 0.2 kg.m^2
Slip:	g = 0.015

Table 7. Parameters of the turbine

Rated Power:	4.5 kWatts
Number of Pole pairs:	P = 2
Blade diameter	R = 3m
Gain:	G = 3.9
The moment of inertia	Jt = 0.00065 kg.m^2
Friction coefficient	ft = 0.017 N.m/s
Air density:	ρ = 1.22 kg/m^3

REFERENCES

Abad Biain Gonzalo. *"Predictive Direct Control Techniques of The Doubly Fed Induction Machine for Wind Energy Generation Applications"*, PhD Thesis (English language), Mondragon Unibersitatea, Spain, 2008.

Abad G., J.Lopez; M.A.Rodrıguez; L.Marroyo and G.Iwanski. Doubly fed induction machine: modeling and control for wind energy generation *IEEE press SPE.* 2011.

Abad Gonzalo and Grzegorz Iwanski. Chapter-10: Properties and Control of a Doubly Fed Induction Machine from the Book: *"Power Electronics for Renewable Energy Systems, Transportation and Industrial Applications"*, EEE Press and John Wiley & Sons Ltd, 2014.

Abdulla Tanvir Aman, Adel Merabet and Rachid Beguenane. Real-Time Control of Active and Reactive Power for Doubly Fed Induction Generator (DFIG)-Based Wind Energy Conversion System, *Energies*, 2015, vol. 8.

Amrane Fayssal and Azeddine Chaiba. A Hybrid Intelligent Control based on DPC for grid connected DFIG with a Fixed Switching Frequency using MPPT Strategy, 4th International Conference on Electrical Engineering, *4th IEEE Conference, ICEE*, 2015.

Amrane Fayssal and Azeddine Chaiba. A Novel Direct Power Control for grid-connected Doubly Fed Induction Generator based on Hybrid Artificial Intelligent Control with Space Vector Modulation, *Rev. Roum. Sci. Techn.– Électrotechn. et Énerg*, 2016b, *vol.* 61.

Amrane Fayssal and Azeddine Chaiba. Type2 Fuzzy Logic Control: Design and Application in Wind Energy Conversion System based on DFIG via Active and Reactive Power Control, Chapter-1, pp.1-35, Title Book: *"Fuzzy Control Systems, Analysis and Performances Evaluation"*, Nova Science Publishers, New York, USA, January 2017a.

Amrane Fayssal, Azeddine Chaiba and Ali Chebabhi. Improvement Performances of Doubly Fed Induction Generator via MPPT Strategy using Model Reference Adaptive Control based on Direct Power

Control with Space Vector Modulation, *Journal of Electrical Engineering,* 2016d, vol. 16.

Amrane Fayssal, Azeddine Chaiba and Saad Mekhilef. High performances of grid-connected DFIG based on direct power control with fixed switching frequency via MPPT strategy using MRAC and Neuro-Fuzzy control, *Journal of power technologies,* 2016c, vol. 96.

Amrane Fayssal, Azeddine Chaiba, Badr-eddine Babes and Saad Mekhilef. Design and Implementation of high Performance Field Oriented Control for Grid-connected Doubly Fed Induction Generator via Hysteresis Rotor Current Controller, *Rev. Roum. Sci. Techn. – Électrotechn. et Énerg.* 2016a, vol. 61.

Behzad Vafakhah. *Multilevel Space Vector PWM for Multilevel Coupled Inductor Inverters,* University of Alberta, Ph.D. Thesis (English language), Canada, 2010.

Bekakra Youcef and Djilani Ben Attous. DFIG sliding mode control fed by back-to-back PWM converter with DC-link voltage control for variable speed wind turbine", *Front. Energy,* 2014, vol. 8.

Brod D. M.and D.W. Novotny. Current control of VSI-PWM inverters, *IEEE Transaction on Industrial Application,* 1985, vol. 21.

Cardenas Roberto, Rubén Pena, Salvador Alepuz and Greg Asher. Overview of Control Systems for the Operation of DFIGs in Wind Energy Applications, *IEEE Transactions on Industrial Electronics,* 2013, vol. 60.

Chaiba Azeddine, Rachid Abdessemed and M. L. Bendaas. Hybrid Intelligent Control based Torque Tracking approach for Doubly Fed Asynchronous Motor (DFAM) drive, *Journal of Electrical Systems,* 2012, vol. 08.

Chaiba Azeddine. A neuro-fuzzy control based torque tracking approach for doubly fed induction generator 4th *IEEE International Conference on Power Engineering Energy and Electrical Drives, Istanbul Turkey,* 2013.

Elmas Cetin, Oguz Ustun and Hasan H. Sayan. A neuro-fuzzy controller for speed control of a permanent magnet synchronous motor drive, *Expert Systems with Applications,* 2008, vol. 34.

Enev S. E. Input-output linearization control of current-fed induction motors with rotor resistance and load torque identification, *1st IEEE ICEIE Hammamet, Tunisia*, 2006.

Fayssal Amrane, Azeddine Chaiba and Bruno Francois. Suitable Power Control based on Type-2 Fuzzy Logic Control for Wind-Turbine DFIG Under Hypo-Synchronous Mode Fed by NPC Converter, *5th International Conference on Electrical Engineering, IEEE Conference*, 2017b.

Gökbulut Muammer, Beşir Dandil and Cafer Bal. A hybrid neuro-fuzzy controller for brushless DC motors, *Artificial Intelligence and Neural Networks*, 2006.

Guo J., X. Cai, and Y. Gong. Decoupled control of active and reactive power for a grid-connected doubly-fed induction generator, *Nanjing, China*, 2008.

Hu S., Y. Kang, D. Li and X. Lin. Nonlinear control strategy for doubly-fed induction generator (DFIG) in wind power controller, *IEEE conference* 2010.

Itsaso Martinez M., Gerardo Tapia, Ana Susperregui and Haritza Camblong. Sliding-Mode Control for DFIG Rotor- and Grid-Side Converters under Unbalanced and Harmonically Distorted Grid Voltage, *IEEE Transactions on Energy Conversion*, 2012, vol. 27.

Jujawa M., Large wind rising, *Renewable Energy World*, 2003, vol. 6.

Kouzi K., M. Nait-Said, M. Hilairet and É. Berthelot. A fuzzy sliding-mode adaptive speed observer for vector control of an induction motor, *IEEE International Conference on Industrial Electronics IECON*, 2008.

Li Shenghu. Power Flow Modeling to Doubly-Fed Induction Generators (DFIGs) Under Power Regulation, *IEEE Transactions on Power Systems*, 2013, vol. 28.

Marchi Rodrigo, Paulo Dainez, Fernando Von Zuben and Edson Bim. A multilayer perceptron controller applied to the direct power control of a doubly fed induction generator, *IEEE Transactions on Sustainable Energy*, 2014, vol. 5.

Mohammadi Jafar, Sadegh Vaez-Zadeh, Saeed Afsharnia, Ehsan Daryabeigi. A Combined Vector and Direct Power Control for DFIG-Based Wind Turbines, *IEEE Transactions on Sustainable Energy,* 2014, vol. 5.

Mullane A., G. Lightbody and R. Yacamini. Modeling of the wind turbine with a doubly-fed induction generator for grid integration studies *IEEE TEC.* 2006, vol. 21.

Mwaniki Julius, Hui Lin, and Zhiyong Dai. A Condensed Introduction to the Doubly Fed Induction Generator Wind Energy Conversion Systems", *Hindawi Journal of Engineering,* 2017.

Shen Baike, Bakari Mwinyiwiwa, Yongzheng Zhang and Boon-Teck Ooi. Sensorless Maximum Power Point Tracking of Wind by DFIG Using Rotor Position Phase Lock Loop (PLL), *IEEE Transactions on Power Electronics,* 2009, vol. 24.

Stemmler H., P. Geggenbach. Configurations of high power voltage source inverter drives, *Proceeding of the 5th European conference on power electronics Brighton,* UK, 1993, vol. 5.

Susperregui A., G. Tapia, I. Zubia and J. X. Ostolaza. Sliding-Mode control of Doubly-Fed Generator for Optimum Power Curve Tracking, *Electronics Letters,* 2010, vol. 46.

Swami Naidu N. K. and Bhim Singh. Experimental Implementation of a Doubly Fed Induction Generator Used for Voltage Regulation at a Remote Location, *IEEE Transactions on Industry Applications,* 2016, vol. 52.

Tohidi Akbar, Hadi Hajieghrary and M. Ani Hsieh. Adaptive Disturbance Rejection Control Scheme for DFIG-Based Wind Turbine: Theory and Experiments, *IEEE Transactions on Industry Applications,* 2016, vol. 52.

Tremblay Etienne, Sergio Atayde and Ambrish Chandra. Comparative Study of Control Strategies for the Doubly Fed Induction Generator in Wind Energy Conversion Systems: A DSP-Based Implementation Approach, *IEEE Transactions on Sustainable Energy,* 2011, vol. 2.

Yao J., H. Li, Y. Liao, and Z. Chen. An improved control strategy of limiting the DC-link voltage fluctuation for a doubly fed induction wind generator, *IEEE Transaction on Power Electronics.* 2008, vol. 23.

Zhang Z.-S., Y. -Z. Sun, J. Lin and G. -J. Li. Coordinated frequency regulation by doubly fed induction generator-based wind power plants, *IET Renewable Power Generation,* 2012, vol. 6.

Zhou Keliang and Danwei Wang. Relationship between Space-Vector Modulation and Three-Phase Carrier-Based PWM: A Comprehensive Analysis, *IEEE Transactions on Industrial Electronics,* 2002, vol. 49.

Zou Yu, Malik E. Elbuluk and Yilmaz Sozer. Stability Analysis of Maximum Power Point Tracking (MPPT) Method in Wind Power Systems, *IEEE Transactions on Industry Applications,* 2013, vol. 49.

BIOGRAPHICAL SKETCHES

Fayssal Amrane

Fayssal Amrane was born in Merouana (Batna), Algeria, in 1989. He received his License and Master degrees in electrical engineering from University of Setif-1, Algeria, in 2010 and 2012, respectively. After that, he worked in the industry for some months. Actually, he is member of the LAS laboratory Setif-1 University. Actually, he is member of the LAS laboratory Setif-1 University as assistant professor. His doctorate dissertation was focused on "Contribution to the Control of Wind Energy Conversion System based on Doubly Fed Induction Generator", this PhD dissertation was collaboration work under PROFAS b+ scholarship program between LAS research Laboratory and L2EP Laboratory from Ecole Centrale de Lille/France. His specific research interests are in the areas of nonlinear controls of Wind Energy Conversion System especially using the DFIG, Intelligent Control, Multilevel Converter, Matrix converter.

Affiliation: Department of Electrical Engineering, University of Setif 1, 19000 Setif, Algeria.

Education:
- PhD in Electrical engineering, University of Setif-1, Algeria, 2018.
- License and Master Degrees in Control of Electrical machines Department of Electrical Engineering, University of Setif-1, 2010 & 2012 respectively.

Research and Professional Experience:

2018-present: Assistant Professor, Faculty of Science and Technology, Department of Electrical Engineering (Module: Power Electronics, Electrical Machines), at Setif-1 University.

2017-2018: Assistant Professor, Faculty of Economic Sciences, Business and Management, Department of Management (Module: Computer Practical Work), Setif-1 University.

Assistant Professor, Faculty of Arts and Languages, Department of Literature and French Language (Module: Computer Practical Work), at Setif-2 University.

2015-2016: Assistant Professor, Faculty of Economic Sciences, Business and Management, Department of Management, Setif-1 University.

His Current Research Interests Include: Wind energy conversion systems, doubly fed induction generator and its controls, nonlinear control, Space vector modulation, Multi-levels converter, Matrix converter, Power quality and Artificial intelligence (*Type-1 fuzzy logic control, Type-2 fuzzy logic control and Neuro-fuzzy control*).

Journal Papers (2016 and 2017):

- Amrane Fayssal, Azeddine Chaiba and Saad Mekhilef: "High performances of grid-connected DFIG based on direct power control with fixed switching frequency via MPPT strategy using MRAC and Neuro-Fuzzy control" *Journal of power technologies,* Vol. 96, n: 1, 2016.

- Amrane Fayssal and Azeddine Chaiba, "Performances of Type-2 Fuzzy Logic Control and Neuro-Fuzzy Control Based on DPC for Grid Connected DFIG with Fixed Switching Frequency", *International Journal of Electrical, Computer, Energetic, Electronic and Communication Engineering,* Vol: 10, n°: 7, 2016.

- Amrane Fayssal and Azeddine CHAIBA, "A Novel Direct Power Control for Grid-Connected Doubly Fed Induction Generator based on Hybrid Artificial Intelligent Control with Space Vector Modulation", *Rev. Roum. Sci. Techn.– Électrotechn. et Énerg* Vol: 61, n°: 3, pp. 263-268, 2016.

- Amrane Fayssal, Azeddine Chaiba and Ali Chebabhi: "Improvement Performances of Doubly Fed Induction Generator via MPPT Strategy using Model Reference Adaptive Control based on Direct Power Control with Space Vector Modulation", *Journal of Electrical Engineering,* Vol: 16, n°: 3, pp. 218-225, 2016.

- Amrane Fayssal, Azeddine Chaiba, Badr Eddine Babes and Saad Mekhilef, "Design and Implementation of High Performance Field Oriented Control for Grid-Connected Doubly Fed Induction Generator via Hysteresis Rotor Current Controller", *Rev. Roum. Sci. Techn.– Électrotechn. et Énerg* Vol: 61, n°: 4, pp. 319-324, 2016.

Chapter in Book (2017):
- Amrane Fayssal and Azeddine Chaiba, "Type2 Fuzzy Logic Control: Design and Application in Wind Energy Conversion System based on DFIG via Active and Reactive Power Control", Chapter-1, pp.1-35, Title Book: "Fuzzy Control Systems, Analysis and Performances Evaluation", Nova Science Publishers, Inc., New York, USA, January 2017. ISBN: 978-1-63485-889-2.

International Conferences in: 2015, 2016 and 2017:
- Amrane Fayssal and Azeddine Chaiba: "Model Reference Adaptive Control for DFIG based on DPC with a Fixed Switching Frequency" *International Electrical Computer Engineering Conference,* IECEC 23-25th May 2015 Setif-Algeria.
- Amrane Fayssal and Azeddine Chaiba: "Direct Power Control for grid-connected DFIG using Fuzzy Logic with a Fixed Switching Frequency" *Conférence Internationale d'Automatique et de la Mécatronique,* CIAM 10-11th Nov 2015 Oran-Algeria.
- Amrane Fayssal and Azeddine Chaiba: "Comparative Study on the performance of Fuzzy-PID and MRAC-PID Controllers based on DPC with SVM for DFIG using MPPT Strategy" *International Conference on Automatic Control, Telecommunication and Signals,* ICATS 16-18th Nov 2015 Annaba-Algeria.
- Amrane Fayssal and Azeddine Chaiba: "A Hybrid Intelligent Control based on DPC for grid-connected DFIG with a Fixed

Switching Frequency using MPPT Strategy" 4th International Conference on Electrical Engineering, *IEEE Conference*, ICEE 13-15th Dec 2015 Boumerdes-Algeria.

- Amrane Fayssal, Azeddine Chaiba and Khaled Eben el-Oualid. Medani: "Neuro-Fuzzy Control based on DPC for grid-connected DFIG with a Fixed Switching Frequency" *International conference on electrical engineering and first workshop on robotics and controls,* 9th CEE 2-4th October 2016 Batna-Algeria.
- Amrane Fayssal, Azeddine Chaiba and Khaled Eben el-Oualid. MEDANI: "Improved Input-Output Linearizing Control using MRAC in Variable Speed DFIG-based on WECS Fed by Three-Level Voltage Source Inverter" *International conference on electrical engineering and first workshop on robotics and controls,* 9th CEE 2-4th October 2016 Batna-Algeria.
- Amrane Fayssal, Azeddine Chaiba and Ali Chebabhi: "Improved Active and Reactive Power Control WECS for grid-connected DFIG using Type-1 and Type-2 Fuzzy Logic Control" *International Conference on Technological Advances in Electrical Engineering,* ICTAEE 24-26th October 2016 Skikda-Algeria.
- Amrane Fayssal, Azeddine Chaiba and Ali Chebabhi: "Robust and Simplified Input-Output Linearizing Control in Variable Speed DFIG using MRAC with Fixed Switching Frequency", *International Conference on Technological Advances in Electrical Engineering,* ICTAEE 24-26th October 2016 Skikda-Algeria.
- Amrane Fayssal, Azeddine Chaiba and Bruno Francois: "Application of Adaptive T2FLC in Stator Active and Reactive power Control WECS based on DFIG via Hypo/Hyper-Synchronous Modes", *4ième Conférence des Jeunes Chercheurs en Génie Electrique,* JCGE, 30 Mai et 1er Juin 2017, Arras, France.
- Amrane Fayssal, Azeddine Chaiba, Bruno Francois and Badr Eddine BABES: "Experimental Design of Stand-alone Field Oriented Control for WECS in Variable Speed DFIG-based on Hysteresis Current Controller" *15th International Conference on*

Electrical Machines, Drives and Power Systems ELMA, 1-3th June 2017 Sofia-Bulgaria.

- Amrane Fayssal, Azeddine Chaiba, Bruno Francois and Badreddine BABES: "Real Time Implementation of Grid-connection control using Robust PLL for WECS in Variable Speed DFIG-based on HCC", *5th International Conference on Electrical Engineering, IEEE Conference*, ICEE 29-31th Oct 2017 Boumerdes-Algeria.
- Amrane Fayssal, Azeddine Chaiba and Bruno Francois: "Suitable Power Control based on Type-2 Fuzzy Logic Control for Wind-Turbine DFIG Under Hypo-Synchronous Mode Fed by NPC Converter" *5th International Conference on Electrical Engineering, IEEE Conference*, ICEE 29-31th Oct 2017 Boumerdes-Algeria.

Azeddine Chaiba

Azeddine Chaiba was born in Ain Zaatout, Algeria, in 1977. He received the PhD, HdR and Pr degrees in electrical engineering from University of Batna, Algeria, in 2010, 2013 and 2018 respectively. He was with Department of Electrical Engineering, University of Setif-1, Algeria until 2017. He is currently an Associate Professor at the Department of Industrial Engineering University of Khenchela, Algeria. His research interest is: Wind energy conversion, Control of doubly fed induction generator (DFIG), Artificial intelligent control, AC-DC-AC converter.

Affiliation: Department of Industrial Engineering University of Khenchela, Algeria.

Education:

- HdR in Electrical engineering, University of Batna, Algeria, 2013.
- PhD in Electrical engineering, University of Batna, Algeria, 2010.
- Msc degree in Control of Electrical machines Department of Electrical Engineering, University of Batna, Algeria, 2004.
- Bsc degree in Electrical network, Department of Electrical Engineering, University of Batna, Algeria, 2001.

Address: Department of Industrial Engineering University of Khenchela.

Research and Professional Experience:

2017-present: Full Professor, Department of Industrial Engineering University of Khenchela

2010-2017: Associate Professor, Department of Electrical Engineering, University of Setif.

2008-2010: Assistant Professor, Department of Electrical Engineering, University of Setif.

2006-2008: Assistant Professor, Department of Electrical Engineering, Batna University.

Journal Papers:

- Chaiba A., R. Abdessemed, and M. L. Bendaas, "A hybrid intelligent control based torque tracking approach for Doubly Fed Asynchronous Motor (DFAM) drive", *Journal of.Electrical Systems,* Vol 9, No.3, pp. 1-13, 2012.
- Chaiba A., R. Abdessemed, and M. L. Bendaas, "A neuro-fuzzy controller for doubly fed asynchronous motor drive", I-manager's *Journal on Electrical Engineering,* Vol. 4, No. 1, July- September, pp. 8-15, 2010.

- Chaiba A., R. Abdessemed, and M. L. Bendaas, "A Torque Tracking Control Algorithm for Doubly-Fed Induction Generator", *Journal of Electrical Engineering Elektrotechnický èasopis, JEEEC,* Vol.59, No.3, pp. 165-168, Slovakia, 2008.
- Chaiba A., R. Abdessemed, M. L. Bendaas, "Control of Torque and Unity stator Side Power Factor of the Doubly-Fed Induction Generator", *International journal of Electrical and Power Engineering* 1 (4), pp. 377-381, 2007.
- Chaiba A., R. Abdessemed, M. L. Bendaas and A. Dendouga, "Performances of Torque Tracking Control for Doubly Fed Asynchronous Motor using PI and Fuzzy Logic Controllers", *Journal of Electrical Engineering, JEE.* Vol.5, N°2, pp25-30, 2005.
- Amrane Fayssal, Azeddine Chaiba and Saad Mekhilef: "High performances of grid-connected DFIG based on direct power control with fixed switching frequency via MPPT strategy using MRAC and Neuro-Fuzzy control" *Journal of power technologies,* Vol. 96, n°: 1, 2016.
- Amrane Fayssal and Azeddine Chaiba, "Performances of Type-2 Fuzzy Logic Control and Neuro-Fuzzy Control Based on DPC for Grid Connected DFIG with Fixed Switching Frequency", *International Journal of Electrical, Computer, Energetic, Electronic and Communication Engineering,* Vol: 10, n°: 7, 2016.
- Amrane Fayssal and Azeddine Chaiba, "A Novel Direct Power Control for Grid-Connected Doubly Fed Induction Generator based on Hybrid Artificial Intelligent Control with Space Vector Modulation", *Rev. Roum. Sci. Techn.– Électrotechn.* et Énerg Vol: 61, n°: 3, pp. 263-268, 2016.
- Amrane Fayssal, Azeddine Chaiba and Ali Chebabhi: "Improvement Performances of Doubly Fed Induction Generator via MPPT Strategy using Model Reference Adaptive Control based on Direct Power Control with Space Vector Modulation", *Journal of Electrical Engineering,* Vol: 16, n°: 3, pp. 218-225, 2016.

- Amrane Fayssal, Azeddine Chaiba, Badr Eddine BABES and Saad Mekhilef, "Design and Implementation of High Performance Field Oriented Control for Grid-Connected Doubly Fed Induction Generator via Hysteresis Rotor Current Controller", *Rev. Roum. Sci. Techn.–* Électrotechn. et Énerg Vol: 61, n°: 4, pp. 319-324, 2016.

Chapter in Book (2017):
- Fayssal Amrane and Azeddine Chaiba, "Type2 Fuzzy Logic Control: Design and Application in Wind Energy Conversion System based on DFIG via Active and Reactive Power Control", Chapter-1, pp.1-35, Title Book: "*Fuzzy Control Systems, Analysis and Performances Evaluation*", Nova Science Publishers, Inc., New York, USA, January 2017. ISBN: 978-1-63485-889-2.

International Conferences:
- Chaiba, "A neuro-fuzzy control based torque tracking approach for doubly fed induction generator", *4th International Conference on power engineering energy and electrical drives,* Istanbul, Turkey,13-17 May, 2013.
- Chaiba, R. Abdessemed, M. L. Bendaas and A. Dendouga, "A Torque Tracking Control Algorithm for Doubly-Fed Induction Machine", *Third IEEE International Conference on ystems,*

Novel I/O-LDC Control Based on ANFIS ... 107

Signals & Devices 'SSD'05', Volume II, march 21-24, 2005, Sousse, Tunisia.

- Chaiba, R. Abdessemed, M. L. Bendaas and A. Dendouga, "Performances of Fuzzy Logic based Torque Tracking Control for Doubly Fed Induction Generator", *First international conference on Electrical Systems PCSE'05*, Oum El-Bouaghi, Algeria, proc. pp 236-241, May 9-11, 2005.
- Chaiba, R. Abdessemed, M. L. Bendaas and A. Dendouga, "Evaluation of the High Performance Vector Controlled Doubly-Fed Induction Generator (DFIG)", *3rd Conference on Electrical Engineering "CEE'04"*, Batna University, proc. pp117-120, 04-06 October, 2004.
- Fayssal Amrane and Azeddine Chaiba: "Model Reference Adaptive Control for DFIG based on DPC with a Fixed Switching Frequency", *International Electrical Computer Engineering Conference*, IECEC 23-25th May 2015 Setif-Algeria.
- Fayssal Amrane and Azeddine Chaiba: "Direct Power Control for grid-connected DFIG using Fuzzy Logic with a Fixed Switching Frequency", *Conférence Internationale d'Automatique et de la Mécatronique*, CIAM 10-11th Nov 2015 Oran-Algeria.
- Fayssal Amrane and Azeddine Chaiba: "Comparative Study on the performance of Fuzzy-PID and MRAC-PID Controllers based on DPC with SVM for DFIG using MPPT Strategy", *International Conference on Automatic Control, Telecommunication and Signals, ICATS* 16-18th Nov 2015 Annaba-Algeria.
- Fayssal Amrane and Azeddine Chaiba: "A Hybrid Intelligent Control based on DPC for grid-connected DFIG with a Fixed Switching Frequency using MPPT Strategy", *4th International Conference on Electrical Engineering, IEEE Conference*, ICEE 13-15th Dec 2015 Boumerdes-Algeria.
- Fayssal Amrane, Azeddine Chaiba and Khaled Eben el-Oualid. MEDANI: "Neuro-Fuzzy Control based on DPC for grid-connected DFIG with a Fixed Switching Frequency", *International*

conference on electrical engineering and first workshop on robotics and controls, 9th CEE 2-4th October 2016 Batna-Algeria.

- Fayssal Amrane, Azeddine Chaiba and Khaled Eben el-Oualid. MEDANI: "Improved Input-Output Linearizing Control using MRAC in Variable Speed DFIG-based on WECS Fed by Three-Level Voltage Source Inverter", *International conference on electrical engineering and first workshop on robotics and controls,* 9th CEE 2-4th October 2016 Batna-Algeria.
- Fayssal Amrane, Azeddine Chaiba and Ali Chebabhi: "Improved Active and Reactive Power Control WECS for grid-connected DFIG using Type-1 and Type-2 Fuzzy Logic Control", *International Conference on Technological Advances in Electrical Engineering,* ICTAEE 24-26th October 2016 Skikda-Algeria.
- Fayssal Amrane, Azeddine Chaiba and Ali Chebabhi: "Robust and Simplified Input-Output Linearizing Control in Variable Speed DFIG using MRAC with Fixed Switching Frequency", *International Conference on Technological Advances in Electrical Engineering,* ICTAEE 24-26th October 2016 Skikda-Algeria.
- Fayssal Amrane, Azeddine Chaiba and Bruno Francois: "Application of Adaptive T2FLC in Stator Active and Reactive power Control WECS based on DFIG via Hypo/Hyper-Synchronous Modes", *4ième Conférence des Jeunes Chercheurs en Génie Electrique, JCGE, 30* Mai et 1er Juin 2017, Arras, France.
- Fayssal Amrane, Azeddine Chaiba, Bruno Francois and Badr Eddine BABES: "Experimental Design of Stand-alone Field Oriented Control for WECS in Variable Speed DFIG-based on Hysteresis Current Controller" *15th International Conference on Electrical Machines, Drives and Power Systems ELMA,* 1-3th June 2017 Sofia-Bulgaria.
- Fayssal Amrane, Azeddine Chaiba, Bruno Francois and Badreddine BABES: "Real Time Implementation of Grid-connection control using Robust PLL for WECS in Variable Speed DFIG-based on HCC", *5th International Conference on Electrical*

Engineering, IEEE Conference, ICEE 29-31th Oct 2017 Boumerdes-Algeria.

- Fayssal Amrane, Azeddine Chaiba and Bruno Francois: "Suitable Power Control based on Type-2 Fuzzy Logic Control for Wind-Turbine DFIG Under Hypo-Synchronous Mode Fed by NPC Converter", *5th International Conference on Electrical Engineering, IEEE Conference,* ICEE 29-31th Oct 2017 Boumerdes-Algeria.

In: Advances in Engineering Research
Editor: Victoria M. Petrova
ISBN: 978-1-53616-092-5
© 2019 Nova Science Publishers, Inc.

Chapter 3

CORROSION EFFECTS ON STEEL REINFORCEMENT

Ch. Apostolopoulos[*] and Arg. Drakakaki
Department of Mechanical Engineering and Aeronautics,
University of Patras, Greece

ABSTRACT

Reinforced concrete is one of the world's most common building materials. Its utility and versatility are achieved through the combination of the best features of concrete and steel. Thanks to concrete a great resistance to compression can be recorded and thanks to steel high ductility and consequently significant resistance to tension can be developed. However, deterioration of such structures, due to ageing, which is owed to environmental factors, is a frequent phenomenon.

Among the main ageing factors though, is corrosion due to chlorides penetration. Mechanical degradation of steel, concrete spalling or even premature failure are some of the main consequences of diffusion of chlorides, which primarily initiate from areas where poorly processed concrete or inadequate concrete thickness can be found. Of course, permeability of concrete, combined with existing surface defects, or even subsistent corrosion damage detected in steel, is responsible for harsher

---

[*] Corresponding Author's Email: charrisa@upatras.gr.

damage, which in long-term can be proved destructive and liable to cause unexpected consequences or situations. Additional factors, which determine significantly the severity of damage, besides the environmental parameters, are the length of the reinforcement, which is exposed to the aggressive conditions -each time- in function of the length of the adjacent areas, which are still protected by concrete, due to the "differential aeration corrosion" phenomenon. Therefore, structural elements with exposed areas of different length, on their reinforcement, are expected to demonstrate uniform level of corrosion damage, which consequently results in differentiated mechanical degradation and unavoidably dissimilar performance against imposed loads.

Analogous parameters and remarks raise several issues, and trigger the scientific community's interest on the laboratory investigation of such topics. Two of the most popular methods, which are used for the laboratory simulation and reproduction of the corrosive conditions, are "salt spray chamber" and "impressed current density" technique. Salt spray chamber tests are carried out in accordance to ASTM B117 specification; however, there are no similar specifications available for impressed current density technique.

Keywords: corrosion, steel reinforcement, parameters, degradation reinforced concrete structures

INTRODUCTION

Reinforced concrete has been one of the most popular structural materials for more than a century. When reinforced concrete structures are located in nonaggressive environments, they usually correspond to the predictions referring to their life expectancy. However, early signs of degradation are exhibited when structures are located in coastal aggressive environments, where physical, chemical or mechanical processes are taking place [1].

One of the most common degradation phenomena of reinforced concrete, which is owed to electrochemical processes, is corrosion due to chlorides penetration. Environmental conditions, quality of materials and constructive parameters are the factors that determine initiation and propagation of the corrosion process. Corrosion reaction requires the existence of an anode, a cathode, an electron pathway and electrolyte (ionic pathway).

During the early years after the construction of a reinforced concrete structure, steel in is in passive conditions and it is protected by a thin layer of oxide which is due to the alkalinity of concrete (pH between 12 to 13) [2-6]. However, presence of humidity, moisture and oxygen in combination with chloride ions, acts as catalyst for corrosion to occur [7-10]. Chlorides penetrate into concrete through the pore network and micro cracks, forming the oxide film over the reinforcing steel and hence, accelerates the reaction of corrosion and concrete deterioration [6, 11-12]. When steel reacts with oxygen, in presence of water or air moisture rust is formed.

Free oxygen and dissolved iron form iron oxide, releasing electrons, which can flow to another part of the metal. As it can be easily understood, because of the electrochemical nature of the reaction, dissolved electrolytes in water aid the reaction. Rust occurs more quickly in saltwater than in pure water, for example.

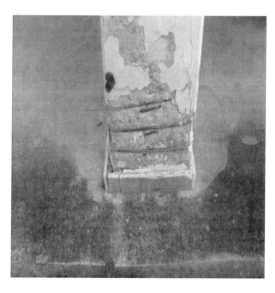

Figure 1. View of a degraded base of a column, exposed to corrosive environment [14].

Steel corrosion in concrete leads to cracking, spalling, and reduction of bond strength, reduction of steel cross section and loss of serviceability. Reinforced concrete undergoing corrosion does not only give the appearance of poor performance, but can in extreme cases, lose its structural integrity [13]. Figure 1 presents a reinforced concrete column affected by corrosion, where crack growth and spalling phenomena are observed.

Corrosion attack can be broadly classified either as uniform or localized. Of course, the consequences of the two forms of attack differ remarkably. Precisely, uniform corrosion effect on steel reinforcement can be estimated as the loss of nominal diameter of the rebar, whereas, localized corrosion, which is more difficult to detect, is more treacherous. The localized attack in combination with the confined rust products, make damage less distinguishable. Besides the optical results, pitting corrosion has a greater impact on ductility of steel reinforcement as well. This is because developing notches are responsible for high stress concentration, in extreme cases, and due to the brittleness of steel, which is owed to release of hydrogen [15]. Nevertheless, classification of corrosion damage is not the only factor determining the effect of the electrochemical phenomenon on steel reinforcement.

As it has been described in recent literature, existing, internal or subcutaneous defects on steel bars [16], or even the differentiated exposed length of steel reinforcement are some additional crucial factors which assign the corrosion damage level [14, 17]. Precisely, existing defects sub serve corrosion propagation and formation of corrosion paths in steel material while different exposed lengths develop dissimilar corrosion levels.

For this reason, further research on the corrosion damage and its consequences was required. The best way for conducting assistive experiments was the development of various accelerated laboratory corrosion test methods. Accelerated tests are useful for the estimation of long-term effects in shorter scales, allowing, for instance the evaluation of long- term degradation effects caused by corrosion. Additionally, they constitute useful tools to establish comparative analysis, which helps identifying how variations in the materials and the corrosion parameters, which have been selected for investigation, may affect their behavior.

The most widespread accelerated corrosion techniques, which have been used for the goals of the presented results, are impressed current density technique and salt spray chamber method. Both methods are used to resemble natural corrosion, in order to evaluate-in laboratory level- the upcoming damage accumulation, which often occurs during the life cycle of a reinforced concrete structure. Moreover, extensive research and artificial approach of natural phenomena often provides further information concerning the developing mechanism.

In the present chapter an effort has been made to describe the corrosion effect on steel reinforcement, as well as to highlight some additional parameters that significantly affect not only the performance of steel reinforcement but also the life expectancy of whole reinforced concrete structures.

CONCRETE AND REINFORCING STEEL

Corrosion of steel reinforcement is one of the major causes of structural deterioration of reinforced concrete structures; as a result of this deterioration, a reinforced concrete structure can incur a considerable reduction in structural serviceability, mechanical performance and safety [18-20]. This is because the ultimate capacity of reinforced concrete structures depends not only on their geometry, reinforcement (both amount and arrangement), materials properties and loading type, but also on their structural scheme [21]. Indicatively, when a reinforced concrete structure undergoes steel corrosion, reduction of the initial steel cross sectional area will occur. High crack widths and high deformations, which are observed in corroded structures, forewarn about the danger of structural collapse. Noteworthy is the fact that, according to recent literature [22-23], level of the existing cracking may significantly affect the corrosion rate.

Although corrosion damage of steel reinforcement, which is owed to environmental conditions, is a time dependent phenomenon, however, it is

116 *Ch. Apostolopoulos and Arg. Drakakaki*

not predicted by the existing regulations. Consequently, the "tools" required for the rehabilitation and the prediction of the residual strength of RC structures via constitutive laws, are not available to engineers, who deal with reconstruction issues [24]. Only recently some efforts were made to propose degradation laws, concerning the remaining strength and ductility of the steel rebars [25-26].

BONDING

Corrosion of steel is one of the most frequent and deterioration issues for reinforced concrete structures. Corrosion factor is responsible for reduction on the cross-sectional area of steel reinforcement, changes in the reinforcing bars mechanical properties [13, 27], cracks and spalling of the cover concrete, and a reduction of the interface between reinforcement and the surrounding concrete. The most severe effect of reinforcement corrosion is the change in bond properties between steel and concrete and the inevitable upcoming bond slip, which has extensively been studied in existing literature [19, 28-30].

Corrosion process results in embedded steel by producing and forming oxides (rust), which occupy 2 to 6 times greater volume of the attacking mass [7], causing tensile stresses in surrounding concrete and, thereafter, leading to gradual concrete cracking development and spalling of the cover concrete [31]. In terms of cracking, longitudinal cracks parallel to steel bars provide easy access of chlorides, moisture and oxygen to a wide area of steel reinforcement, a fact which inevitably leads to acceleration and further development of corrosion. In case of transverse cracks, however, the cathode area is situated between the cracks, where moisture and oxygen have to reach the embedded steel through sound concrete in order to enable the corrosion process [32]. As far as crack width is concerned, in accordance with Eurocode 2 [33], the appearance of cracks, with limited width does not necessarily imply a lack of serviceability or durability of reinforced concrete structures. Nonetheless, in practice, when cracks develop in regions surrounding the reinforcing bars, the force transfer is affected. In cases when this fact is not taken into consideration, bonding capacity is usually overestimated [34].

In existing literature, extended investigation has been conducted on the corrosion effect on bond between steel reinforcement and concrete [35-38]. Moreover, studies by Andrade et al. [39] and Tahershamsi [40] provide information and correlation results of the surface crack width with the corrosion level and the degradation of the bond strength. Precisely, bonding between concrete and steel is substantially influenced by the surface characteristic of steel reinforcement and the bends/hooks which provide anchorage of the steel within the concrete. Given that reinforcement ribs contribute to bonding, loosening of the ribs on the corroded steel's surface (as a result of corrosion) must have impaired the gripping of the corroded steel bar within the concrete and thus giving rise to a weakened bond between concrete and the corroded steel, as mentioned for instance by BRE (2000) [41-42].

A number of additional factors affecting mechanical degradation of bonding, such as ribs geometry, surface profile of rebars, concrete composition, compressive strength of concrete, and cover thickness has been studied in recent literature as well [34,38]. Lin et al. [43] indicate in his study the significant effect of cover thickness and amount of transverse reinforcement on induced surface cracking and on bond strength between concrete and corroded rebars. Unobstructed and prolonged environmental action in reinforced concrete elements of structures are common and bring significant corrosion damage in reinforcement steel and extreme bond strength loss, that facilitates the relative slippage between concrete and steel bar, in such a manner that the structural element demonstrates an inadequate resistance.

CORROSION PROCESS ON STEEL REINFORCEMENT

Corrosion of steel, due to chloride penetration, is an oxidation process, followed by the breakdown of the passive film of the steel. Before depassivation of reinforcement a surface layer of ferric oxide covers and protects the steel in concrete. Upon this layer being damaged or depassivated, corrosion of reinforcement in concrete can be activated. Depassivation of the protective layer can occur by reaching a critical threshold concentration of free chloride ions near the surface of the reinforcement bars.

Upon depassivation has occurred, corrosion of steel reinforcement initiates. The electrochemical reactions that occur during the corrosion process are presented as follows:

Anodic dissolution or oxidation of iron going into aqueous (water) solution:

$$2Fe \rightarrow 2Fe^{2+} + 4e^-$$

Cathodic reduction of oxygen that is dissolved into water also occurs:

$$O2 + 2H_2O + 4e^- \rightarrow 4OH^-$$

The iron ion and the hydroxide ion react to form iron hydroxide:

$$2Fe^{2+} + 4OH^- \rightarrow 2Fe(OH)_2$$

The iron oxide reacts with oxygen to yield red rust, $Fe_2O_3.H_2O$

In the initial phase corrosion cracking does not happen directly on the surface of the concrete structure, but it only shows up when corrosion product reaches its threshold value. Before cracking formation Cl^- penetration initiates via diffusion. Whereas, after formation of the first corrosion cracks on concrete surface, a major inflow of chloride ions through cracks can be noticed, which is responsible for the increase of the corrosion rate. At the same time a lower diffusion degree can be recorded. Acceleration of the corrosion process constitutes a severe menace for safety and structural integrity of the reinforced concrete structures [44].

Corrosion manifests in various types, the most common of which are uniform and pitting corrosion. The former type is more common in bare steel bars, whereas the latter, which is a localized form of corrosion that

Corrosion Effects on Steel Reinforcement 119

produces small holes and propagates deep inside the material, is more common in embedded steel. As it has already been mentioned, it is considered to have greater impact than uniform corrosion, due to its unanticipated occurrence and unpredictable propagation rate. Corrosion rate of pitting corrosion is much faster than general corrosion [45]. Its aggressiveness is owed to the very small size and high penetration ability of chloride ion through passive layer [46-47].

LABORATORY INVESTIGATION

Although corrosion is of major importance for reinforced concrete structures, just during the last two decades the phenomenon managed to capture scientific community's interest. Consequently, this is the reason for the belayed inclusion of the corrosion factor in the Standards referring to designing and rehabilitation techniques. False use of effective values of yield strength and energy stocks, instead of the nominal values defined by international regulations and the willing ignorance of the incidental lack of bonding between steel and concrete after some corrosion level, may result in overestimation of the residual true load- bearing capacity of the structure [48].

For this reason, further research on the form of corrosion damage and its consequences was necessary. Accelerated laboratory corrosion testing can be useful for the establishment of comparative analysis [49].

Between the most widespread accelerated corrosion techniques, which have been used for the goals of many studies, are: salt spray chamber and impressed current density technique.

SALT SPRAY CHAMBER

Salt spray chamber constitutes the oldest method for laboratory corrosion testing. This type of test offers numerous advantages, the most important of which is easy management and low cost. Additionally, there

120 *Ch. Apostolopoulos and Arg. Drakakaki*

is a limited number of standards dedicated to this technique, resulting in a widely known framework.

ASTM B117 [50] specification covers every aspect of the apparatus configuration, procedure and conditions required to create and maintain a salt spray (fog) testing environment. The selection of such a procedure for corroding the specimens, relies on the fact that the salt spray environment lies qualitatively closer to the natural coastal (rich in chlorides) conditions than any other accelerated laboratory corrosion test. In principle, the testing apparatus consists of a closed chamber in which a salted solution atomized by means of a nozzle, produces a corrosive environment of dense saline fog. The most common salt solution used, is prepared by dissolving 5 parts by mass of sodium chloride (NaCl) into 95 parts of distilled water (pH range 6.5–7.2). Temperature inside the salt spray chamber maintains equal to 35°C (+1.1–1.7)°C [51].

IMPRESSED CURRENT DENSITY TECHNIQUE

Impressed current technique is one of the most popular methods used for accelerated laboratory corrosion. This technique is in accordance with ASTM standards, although all the individual parameters needed to create a standard corrosive environment have not been determined yet. One advantage over other accelerated techniques is the ability to control the rate of corrosion, which usually varies due to changes in the resistivity, oxygen concentration, and temperature [52]. An accelerated corrosion test by the impressed current technique is confirmed to be a valid method to study the corrosion process of steel in concrete, and its effects on the damage of concrete cover [53]. The scientific justification for accelerating corrosion using an impressed current is strong, given that the initiation period required for depassivation from years to days gets dramatically reduced. Additionally, the desired rate of corrosion is applied, without downgrading the quality of the corrosion products formed [54].

Auyeung et al. [55] have found that theoretical corrosion mass loss and actual corrosion mass loss are not the same because of various factors,

Corrosion Effects on Steel Reinforcement

such as need for electrical energy to initiate the corrosion, resistivity of concrete, composition of the bar, electrical properties of minerals in concrete, etc. El-Maaddawi et al. [56] revealed that varying the applied current density level to get different degrees of induced corrosion during the same time period may have other effects, which may lead to misinterpretation of test results. This conclusion was confirmed by Apostolopoulos and Drakakaki as well [14].

Impressed current technique works by applying a constant current from a DC source to the tested steel to induce significant corrosion in a short period of time [52]. After applying the current for a given duration, the degree of induced corrosion can be determined theoretically using Faraday's law (Equation 1):

$$M_{th} = \frac{WI_{app}T}{F}$$ (1)

where Mth = theoretical mass of rust per unit surface area of the bar (g/cm2); W = equivalent weight of steel which is taken as the ratio of atomic weight of iron to the valency of iron (27.925 g); Iapp = applied current density (Amp/cm2); T = duration of induced corrosion (sec); and F = Faraday's constant (96487 C mol-1).

The percentage of actual mass loss of the specimens tested can be estimated with the use of Equation 2.

$$\Delta m\ (\%) = \frac{m_i - m_f}{m_i} * 100\%$$ (2)

Where mi and mf are the initial and the final mass measurements respectively.

In Figures 2 and 3 is presented an indicative automatic system (found in Laboratory of Technology & Strength of Materials of University of Patras), which has been used for the performance of electrochemical corrosion tests. The specific system enables implementation of different corrosive conditions, in terms of impressed current density and ponding cyclic corrosion (wet/dry duration).

Figure 2. An automatic system organized for the performance electrochemical corrosion tests, using impressed current technique.

Figure 3. Control system used in an automatic system for the performance of electrochemical corrosion tests.

EXPOSURE CONDITIONS/CYCLES (WET, WET/DRY)

As it was mentioned to the previous paragraph, Faraday's law estimates the mass loss rates under constant current, which means in wet conditions. Nevertheless, in real environmental conditions there are both wet and dry stages. Consequently, in order to achieve a better approach to the environmental conditions for both methods, a severe exposure environment of wetting/drying (chloride ponding) is required.

Such a testing regime simulates the chloride exposure of marine structures under splash and tidal zones [57]. In reality, structures are subjected to wet and dry periods, rather than a constant relative humidity. Thus, corrosion arrest, as the result of drying, is important in determining the durability of the exposed structures. Chloride ponding application is in agreement with the simulating corrosion methods used in existing studies as well [28, 58-59] and ensures conditions which are qualitatively closer to the natural. This is because during wetting, chloride solution penetrates a layer of the material; during the drying stage, the evaporation front moves inwards and takes some of the chloride with it [58]. It is deducted in theory that atmospheric corrosion rate of metals can be accelerated by increasing the frequency of wet-dry cycling.

Ponding procedure is more important for embedded specimens. However, given the need for comparisons, concerning the type of corrosion damage, among bare and embedded specimens, corrosion procedure should be common for both categories.

Finally, for the case of impressed current density technique, it is proposed that the specimens should be fully immersed in the chloride solution, during the "wet" process, in contrast to experiments described in existing literature, that deal with either partially immersed specimens [60-61] or with the use of soaked fabric. This is proposed because full immersion of the specimens can guarantee constant conditions and consequently results of high quality and accuracy, given that repeatability is feasible, since laboratory conditions (humidity, temperature, and local conditions) cannot affect the test conditions.

FACTORS AFFECTING CORROSION RATE AND PROPAGATION

Corrosion of steel reinforcement in splash and tidal zones contributes to the dramatic reduction of plastic deformation and therefore diminishes the expectations for the development of plastic hinges in the corroded regions of the structures, especially under intense seismic events. However, it should be noted that unexpected intense earthquakes in combination with the non-uniformity of corrosion development on the steel reinforcement -at different zones- may trigger the formation of plastic hinges at non-expected regions of the RC elements as well [62]. Nevertheless, seismic design of the RC structures is mainly based on the assumption of the plastic hinges occurring at the two ends of the elements (Figure 4).

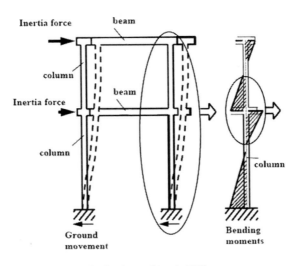

Figure 4. A frame structure under horizontal loads [17].

In an effort to define the reasons that lead to such phenomena, internal structure of steel bars should be examined.

EXISTING DEFECTS

As it is widely known, dual phase high performance steel, which is made of scrap metal, shows an external high strength zone (martensitic phase) and a softer core (ferrite-perlite phase). Beyond these two obvious phases, there is a transition zone called bainite phase. The high mechanical performance of dual phase steel comes from the combination of the mechanical properties of each individual phase, where the increased strength properties are credited to the presence of the outer martensitic zone, and the increased ductility to the ferito-perlitic core. Due to the coexistence of the different phases, an assumption could be made, that the continuity and the coherence of the structure of the specific type of steel is taken for granted. However, this anticipation is not always satisfied, since given the variety of the phases, there are distinct areas of different types of crystallic structures and consequently different types of mechanical behavior on each phase of the material. A proof for this assumption is given in Figure 5. Figure 5 presents the findings of a laboratory investigation, concerning the mechanical behavior of the transition zone of a B500c steel bar sample, after the imposition of eccentric compressive load pointed at the martensitic skin of a cross section on the kinematic behavior of the gauge. In reference to this fact, Figures 5 present various point of views, such as the surface of the slip area and the lower level of the core. Rupture of the two phases and slide lines can also be observed [16].

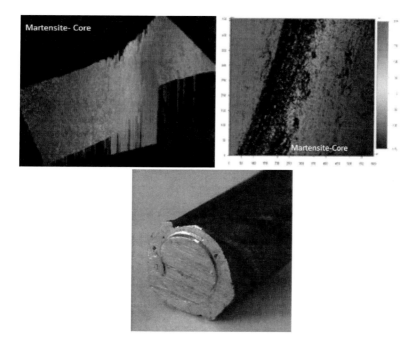

Figure 5. Veeco Device shows the detachment between Martensite–Core (about 400μm) [16].

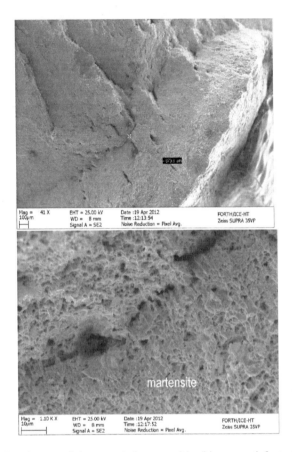

Figure 6. The detachment of the corroded martensitic skin can satisfactorily explain the formation of the "knee" in the elastic region. Focus on the boundary between the martensite and the core of the cross section [63].

In order to confirm further the above-mentioned findings, in recent papers, SEM analyses were performed. In particular, Figures 6 clearly show a localized detachment in the interface of the martensite and the internal core of the precorroded material. In this case, according to the corresponding paper, the crack is located in a distance of approximately 700 μm from the external surface of the steel bar which coincides with the average thickness of the martensitic cortex in dual-phase steel B500c with a nominal 10mm diameter [63].

Moreover, frequently, final products suffer from surface and internal defects, which deteriorate the final product quality. These defects can be formed within steel microstructure during solidification of steel [64]. Some typical examples are voids, cracks, inclusions or other imperfections.

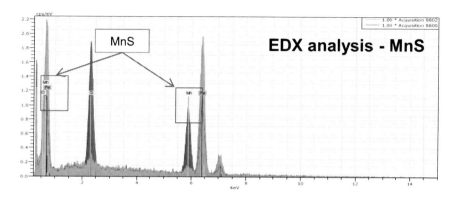

Figure 7. Extended sulfide development in the outer area of the material [16].

Additionally, it is a common knowledge that not all steel bar classes have the same chemical composition and this is something that has been proved by numerous chemical analyses. Specifically, presence of chemical compounds of MnS and FeS or (Mn, Fe) S, as components of steel, has an influence on microstructure of the material (Figure 7). From the viewpoint of fracture mechanics, non-metallic inclusions are equivalent to small defects or cracks that can generate stresses within the surrounding matrix [65-66]. The impact of this behavior of sulfides, as stress concentrators, depends on their size, position and shape, as well as on their ability to bond with the matrix material [66-68]. For example, large sulfides usually result in poor mechanical properties, and non-spherical sulfides are responsible for certain anisotropic properties due to their non determined shape, resulting in the diminishment of the materials performance.

Additionally, existence of chemical sulfide compounds that are generated in steel at the stage of production, greatly affects the behavior of the material against corrosion. Recent studies [70] have proved that among two different steel bar categories, the one with the lower S content -which means and lower sulphide content- demonstrate better corrosion resistance. This is because, as it has been shown, MnS inclusions adversely affect the nature of passive film resulting to accelerated corrosion and pit formation

[16]. Investigation of the size effect of the inclusion on the level of the stress concentration reveals a critical size of the inclusion where the stress concentration does not increase with the growth of the inclusion for a given applied load and boundary conditions. However, during the violent loading, recycles may cause coalescences of critical importance for the steel, due to their proximity, forcing a multiple cracking phenomenon [69]. In addition, voids present a great corrosion risk as well, when they are partially filled with pore solution [70]. Synergy of existing internal and external defects results in corrosion paths creation (Figure 8).

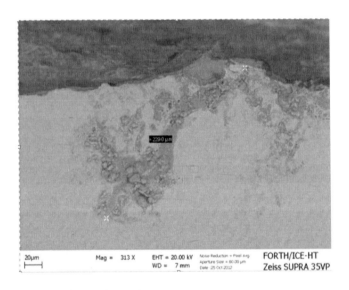

Figure 8. Formation of corrosion paths.

Of course, corrosion mechanism is activated easier in aggressive conditions. According to recent literature, [16, 28, 63, 71], corrosive environment is responsible for the developing surface damage assisted by pits, as well as for the damage recorded within the outer surface, due to sulfide existence on the martensite area. This fact results in the extension of the damage in depth. Additionally, such defects (oxides and sulfides), as well as intergranular corrosion phenomena, not only constitute very serious degradation factors for the material but they also have impact on its mechanical performance. This is because rapid depletion of the ductility, or

Corrosion Effects on Steel Reinforcement 129

even failure, may occur in high strength and ductility dual phase steel bars, due to the combination of interior and exterior damage phenomena under strong stresses [72-73].

However, internal defects are not the only factor affecting internal structure of steel. Another parameter related to the structural integrity of the RC structures is the length of the surface which is exposed to corrosion.

EXPOSED LENGTH

According to Angst U.M. [58] chloride concentration at the steel surface needed to initiate corrosion decreases with increasing length of the reinforced concrete beams exposed to chlorides. To describe the possibilities of concentration of a critical percentage of chlorides in long specimens against short specimens, Angst [58] used in his study equation (3).

$$p_L = 1 - (1 - p_M)^k \tag{3}$$

Where: p_L and p_M are the corrosion possibilities of a long and a short specimen respectively and k is the ratio of the big length to the shorter one.

It is suggested that the size effect can be explained by inhomogeneities found at the steel surface. Consequently, increasing the specimen size will thus increase the probability for the presence of conditions favoring corrosion initiation at lower chloride concentrations. These may include zones of particularly corrosion-susceptible steel metallurgy (for example, accumulations of inclusions) or areas where the concrete adjacent to the steel offers low corrosion protection [58].

A further explanation that can be met in existing literature is that specimens with the same exposed length demonstrate different resistance against corrosion, when their total length varies, because corrosion process is in progress not only on the exposed area but also on the protected area. Figure 9 gives an indicative graph, depicting the different mass loss results recorded, after the same exposure periods, for specimens of short and long exposed length [14].

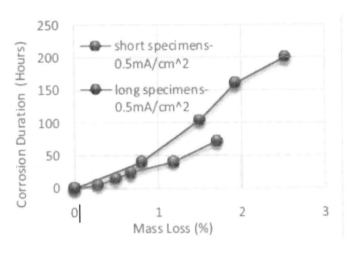

Figure 9. Graphical presentation and comparison of two series of specimens (short and long), of the same diameter Φ12, that were corroded under 0.5mA/cm² [14].

The reason for the dissimilar corrosion rate is the differential aeration reported on the two sections. Precisely, differential aeration corrosion is a type of corrosion that occurs when oxygen concentrations vary across a metal's surface. The varying concentration of oxygen creates an anode and a cathode on the metal's surface. Oxidation then occurs because an anode and a cathode have been established on the surface. In differential aeration corrosion, the area with the higher oxygen concentration becomes the cathode. The area with the lower oxygen concentration becomes the anode. Consequently, the portion of the metal that has the lower oxygen concentration is the portion subject to corrosion. The anodic dissolution rate depends solely upon the potential difference across the electrolyte-metal interface [74].

Such phenomena are not usually taken into consideration for neither for design, nor for rehabilitation of existing structures. Nevertheless, subcutaneous damage may trigger severe issues in terms of mechanical performance and structural integrity of reinforced concrete structures.

CONSEQUENCES OF CORROSION ON THE MECHANICAL BEHAVIOR OF STEEL REINFORCEMENT

Safety of structures is generally related to the expected service life, according to established standards and methods. However, there are plenty of factors that may be responsible for the loss of structural integrity and the reduction of residual life of a structure. It is widely known that harsh coastal environment, which is rich in chlorides, is likely to have detrimental results on the mechanical performance of the reinforced concrete structures. Concrete structures are exposed to chlorides, not only in marine environments, but also in many regions of the world, where deicing salts are used on roads, as a means to improve traffic safety during winter season. Being a porous material, concrete permits penetration of chloride ions through the liquid phase contained in the pore system. Once the sufficiency of high chloride concentration is reached at the reinforcing steel surface, initiation of localized corrosion is possible [61]. According to Graeff A.G. et al. [49], reinforcement corrosion is known as a destructive interaction with the environment, which leads to damage reaction of chemical or electrochemical nature, associated to deterioration processes (both physical and mechanical).

However, besides the fact that corrosion is of major importance for reinforced concrete structures, just a few years ago the phenomenon managed to capture scientific community's interest. Consequently, corrosion factor has not been included in the Standards neither for designing, nor for rehabilitation. False use of effective values of Rp and W, instead of the nominal values defined by international regulations, may result in overestimation of the residual true load- bearing capacity of the structure [49].

The cases of the afore-mentioned underestimation of corrosion damage may frequently be responsible for unexpected damage, especially in critical sites. A typical example of such sites, are the ends of basic structural elements (Figure 1), such as columns or beams. At those areas, cross section is not capable of resisting any additional moment but may maintain the already received moment for some amount of rotation. Consequently, according to the seismic design of RC structures, there is an assumption that plastic hinges will occur, mainly during strong seismic events. However, higher degradation of steel raises questions about reliability of the initial design.

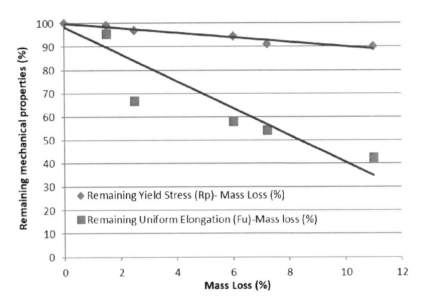

Figure 10. Remaining yield stress and uniform elongation of corroded bare steel samples, in reference to their corresponding mass loss [13].

Besides, it is widely known that corrosion of steel reinforcement in splash and tidal zones contributes to dramatic reduction of plastic deformation as well as to reduction of the load bearing capacity of steel reinforcement, especially under intense seismic events, due to reduction of the nominal diameter [85]. Drop of both yield stress and uniform deformation up to the upcoming failure, has been explicitly discussed in recent literature as well [13], highlighting that aged structures, may be less safe than expected. In Figure 10 are given two lines depicting an indicative drop of the main mechanical properties (yield stress and uniform deformation) of corroded bare steel samples, in reference to the corresponding mass loss, after their exposure to a salt spray chamber. Similar results are presented in Figure 11, referring to embedded specimens of existing aged reinforced concrete structures [13].

Additional loss of material weakens further the material and this is an issue which can easily confute repair and rehabilitation techniques, given that the end of the service lifetime could be earlier than when a crack in the concrete cover is observed. For this reason, further investigation is required.

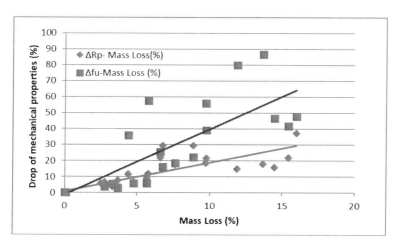

Figure 11. Drop of yield stress and uniform elongation of corroded embedded steel samples, in reference to their mass loss [13].

REFERENCES

[1] Liberati E.A.P., Nogueira C.G., Leonel E.D., Chateauneuf A., 2016. Chapter 5- Failure analysis of reinforced concrete structures subjected to chloride penetration and reinforcements corrosion, *Handbook of Materials Failure Analysis with Case Studies from the Chemicals*, Concrete and Power Industries, 93-121.

[2] CIP 25- corrosion of steel in concrete. Technical information prepared By NRMCA (National Ready Mixed Concrete association).

[3] Mohammed T.U., Otsuki N., Hamada H. 2003. "Corrosion of Steel Bars in Cracked Concrete under Marine Environment". *ASCE, Journal of Material in civil Engineering* 15: 460-469.

[4] Ormellese M., Berra M., Bolzoni F., Pastore T. 2006. "Corrosion inhibitors for chlorides induced corrosion in reinforced concrete structures". *Cement and Concrete Research* 36: 536-547.

[5] Vaysburd A.M., Emmons P.H. 2004. "Corrosion inhibitors and other protective systems in concrete repair: concepts or misconcepts". *Cement & Concrete Composites* 26: 255-263.

[6] Quraishi M.A., Nayak D.K., Kumar R. and Kumar V. 2017. "Corrosion of Reinforced Steel in Concrete and its Control: An overview," *Journal of Steel Structures & Construction, 3*(1).

[7] Isgor O.B., Razaqpur A.G. 2006. Modelling steel corrosion in concrete structures. *Materials and Structures* 39: 291-302.

[8] Gaidis J.M. 2004. Chemistry of corrosion Inhibitors. *Cement and concrete composite* 26: 181-189.

[9] Mullick D.A.K. 2004. Corrosion of reinforcement in concrete-an interactive durability problem. *The Indian Concrete Journal.* 74: 168-175.

[10] Quraishi M.A., Kumar V., Abhilash P.P., Singh B.N. (2011). Calcium Stearate: A Green Corrosion Inhibitor for Steel in Concrete Environment. *J Mater Environ Sci* 2: 365.

[11] Chung L., Kim J.H.J., Yi S.T. 2008. Bond strength prediction for reinforced concrete members with highly corroded reinforcing bars. *Cement & Concrete Composites* 30: 603-611.

[12] Skoglund P., Silfwerbrand J., Holmgren J., Tragardh J. 2008. Chloride redistribution and reinforcement corrosion in the interfacial region between substrate and repair concrete a laboratory study. *Rilem Journals, Materials and Structures* 41: 1001-1014.

[13] Apostolopoulos Ch., Papadakis V. 2007. Consequences of steel corrosion on the ductility properties of reinforcement bar, *Construction and Building Materials, 22*(12):2316-2324.

[14] Apostolopoulos C.A. and Drakakaki A. 2018. "Correlation Between the Electrochemical Corrosion Parameters and the Corrosion Damage, on B500c Dual-Phase Steel," *Innovations in Corrosion and Materials Science*, 8:42-52.

Corrosion Effects on Steel Reinforcement 135

[15] Cairns J., Plizzari G., Du Y., Law D. and Franzoni Ch., 2005. " Mechanical Properties of Corrosion- Damaged Reinforcement," *ACI Materials Journal*, 102 (4):256-264.

[16] Apostolopoulos, Ch., Drakakaki, Arg., Apostolopoulos, Alk., Matikas, T., Rudskoi, A., Kodzhaspirov, G. 2017. Characteristic defects-corrosion damage and mechanical behavior of dual-phase rebar, *Materials Physics and Mechanics*, 30:1-19.

[17] Drakakaki, Arg., Apostolopoulos. Ch. 2018. "The size effect of rebars, on the structural integrity of reinforced concrete structures, which are exposed to corrosive environments," *MATEC Web of Conferences 188:03009.*

[18] Fernandez I., Herrador M., Mari A. and Bairan J. 2018. "Ultimate Capacity of Corroded Statically Indeterminate Reinforced Concrete Members," *International Journal of Concrete Structures and Materials*, 12 (75).

[19] Almusallam A.A. 2001. "Effect of degree of corrosion on the properties of reinforcing steel bars," *Constr. Build Material*, 15: 361-8.

[20] Apostolopoulos Ch. 2009. "The influence of corrosion and cross-section diameter on the mechanical properties of B500c steel," *Journal of Materials Engineering and Performance*, 18:190-195.

[21] Fernandez I., Herrador M., Mari A., Bairan J.M. 2016. "Structural effects of steel reinforcement corrosion on statically indeterminate reinforced concrete members," *Materials and Structures*, 49 (12).

[22] Malumbela, G., Alexander, M., Moyo, P. 2009. "Steel corrosion on RC structures under sustained service loads—a critical review". *Eng Struct* 31:2518–2525.

[23] Malumbela, G., Moyo, P., Alexander, M.2009. "Behaviour of RC beams corroded under sustained service loads," *Constr Build Mater* 23:3346–3351.

[24] Tastani S.P. and Pantazopoulou S.J. 2005. "Recovery of seismic resistance in corrosion-damaged reinforced concrete through FRP Jacketing." invited paper in *International Journal of Materials and Product Technology*, 23(¾): 389 -415.

[25] Apostolopoulos Ch., Drakakaki Arg., Basdeki M. and Apostolopoulos Alk. 2018. "Degradation Laws of Mechanical Properties of Corroded Steel Bar of Existing Structures on Coastal Areas: Natural and Anthropogenic Hazards and Sustainable Preservation," *10th International Symposium on the Conservation of Monuments in the Mediterranean Basin*, DOI: 10.1007/978-3-319-78093-1_15.

[26] Imperatore S., Rinaldi Z., Drago C. 2017. "Degradation relationships for the mechanical properties of corroded steel rebars," *Construction and Building Materials*, 148: 219-230.

[27] Fernandez I., Bairán J.M., Marí A.R. 2015. "Corrosion effects on the mechanical properties of reinforcing steel bars. Fatigue and σ–ε behavior," *Constr Build Mater* 101:772–783. doi: 10.1016/j.conbuild mat.2015.10.139.

[28] Apostolopoulos, C.A., Demis S., Papadakis, V.G.2013. "Chloride-induced corrosion of steel reinforcement – mechanical performance and pit depth analysis," *Constr.Build.Mater.*, 38:139–46.

[29] Papadopoulos, M.D., Apostolopoulos, C.A., Zervaki, A.D., Haidemenopoulos, G.N. 2011. "Corrosion of exposed rebars, associated mechanical degradation and correlation with accelerated corrosion tests," *Jour. Constr. Build Mater*, 25: 3367-3374.

[30] Imperatore S., Rinaldi Z. 2008. "Mechanical behavior of corroded rebars and influence on the structural response of RC elements," *Concrete Repair, Rehabilitation and Retrofitting II*, Cape Town, South Africa.

[31] Elbusaefi, A. 2014. *"The effect of steel bar corrosion on the bond strength of concrete manufactured with cement replacement materials,"* Ph.D. thesis, School of Engineering, Cardiff University, Cardiff, Wales.

[32] Shaikh F.U.A. 2018. "Effect of Cracking on Corrosion of Steel in Concrete," *Internationla Journal of Concrete Structures and Materials*, 12 (3).

Corrosion Effects on Steel Reinforcement 137

[33] EN-1992-1-1, Eurocode 2–Design of Concrete Structures. Part 1–1: *General Rules and Rules for Buildings*, Comité Européen de Normalisation (CEN), Brussels, Belgium, 2008.

[34] Apostolopoulos Ch., Koulouris K., Apostolopoulos Alk.2019. "Correlation of Surface Cracks of Concrete due to Corrosion and Bond Strength (between Steel Bar and Concrete)," *Advances in Civil Engineering*, 2019, Article ID: 3438743.

[35] Tastani S. and Pantazopoulou S. 2009. "Direct tension pullout bond test: experimental results," *Journal of Structural Engineering*, 136(6): 731–743.

[36] Desnerck P., Lees, J. M. and Morley, C. T. 2015. "Bond behaviour of reinforcing bars in cracked concrete," *Construction and Building Materials*, 94: 126–136.

[37] Fang C., Lundgren K., Chen L., and Zhu C. 2004. "Corrosion influence on bond in reinforced concrete," *Cement and Concrete Research*, 34(11): 2159–2167.

[38] Li C. Q. and Zheng J. J. 2005. "Propagation of reinforcement corrosion in concrete and its effects on structural deterioration," *Magazine of Concrete Research*, 57(5): 261–271.

[39] Andrade C., Cesetti A., Mancini G. and Tondolo F. 2016. "Estimating corrosion attack in reinforced concrete by means of crack opening," *Structural Concrete*, 17(4): 533– 540.

[40] Tahershamsi M., Fernandez I., Lundgren K., and Zandi K.2017. "Investigating correlations between crack width, corrosion level and anchorage capacity," *Structure and Infrastructure Engineering*, 13 (10):1294–1307.

[41] BRE (2000). *Corrosion of steel in concrete: Investigation and assessment.* Digest 444-Part 2. BRE.

[42] Adukpo E., Oteng-Seifah S., Manu P., Solomon-Aych K., *The Effect of Corrosion on the Strength of Steel Reinforcement and Reinforced Concrete*, 166-175.

[43] Lin H., Zhao Y., O˘zbolt J., and Hans-Wolf R. 2017. "Bond strength evaluation of corroded steel bars via the surface crack width induced by reinforcement corrosion," *Engineering Structures*, 152: 506–522.

[44] Maruya T., Takeda H., Horiguchi K., Koyama S., Hsu K. L. 2007. "Simulation of Steel Corrosion in Concrete Based on the Model of Macro-Cell Corrosion Circuit" *J. Adv. Concrete Technol.*, 6:343–362.

[45] Darowicki K. 2004. "Evaluation of pitting corrosion by means of dynamic electrochemical impedance spectroscopy" Science Direct, *Electrochimica Acta*, 49(2004):2909-2918.

[46] Standard Guide G48-92, *Annual Book of ASTM Standards*, Philadelphia PA, ASTM-1994.

[47] Rustandi A., Nuradityatama M., Rendi F., Setiawan S. 2017. "The use of Cyclic Polarization Method for Corrosion Resistance Evaluation of Austenitic Stainless Steel 304L and 316L in Aqueous Sodium Chloride Solution," *International Journal of Mechanical Engineering and Robotics Research*, 6 (6): 512-518.

[48] Apostolopoulos C. A., Pasialis V. P. 2008. "Use of quality indices in comparison of corroded technical steel bars B500c and S500s, on their mechanical performance basis," *Construction and Building Materials*, 22:2325-2334.

[49] Graeff A. G., Pinto da Silva Filha L. C. 2008. "Analysis of rebar cross sectional area loss by reinforced concrete corrosion," in *11th International Conference on Durability of Building Materials and Components*, Istanbul, Turkey.

[50] ASTM Standard B117. *Standard Practice for Operating Slat Spray (Fog) Apparatus.* ASTM.

[51] Drakakaki Arg., Apostolopoulos Ch., Katsaounis Al., Hasa B. 2017. "Corrosion resistance and mechanical characteristics of dual phase steel B500c, after shot blasting processes," *International Journal of Structural Integrity*, 8(5).

[52] Ahmad Sh. 2009. "Techniques for inducing accelerated corrosion of steel in concrete," *Arabian Journal for Science and Engineering*, 34 (2C):95-104.

[53] Care S. and Raharinaivo A. 2007. "Influence of Impressed Current on the Initiation of Damage in Reinforced Mortar due to Corrosion

Corrosion Effects on Steel Reinforcement

of Embedded Steel," *Cement and Concrete Research*, 37: 1598–1612.

[54] Austin, S. A., Lyons, R., Ing, M. J. 2004. "Electrochemical Behavior of Steel-Reinforced Concrete during Accelerated Corrosion Testing," *Corrosion*, 60: 203–212.

[55] Auyeung Y., Balaguru P., Chung L. 2000. "Bond Behavior of Corroded Reinforcement Bars," *ACI Materials Journal*, 97: 214–220.

[56] El Maaddawy T. A. and Soudki, K. A. 2003. "Effectiveness of Impressed Current Technique to Simulate Corrosion of Steel Reinforcement in Concrete," *ASCE Journal of Materials in Civil Engineering*, 15: 41–47.

[57] Ma Q., Nanukuttan S. V., Basheer P. A. M., Bai, Y. and Yang, C. 2016, "Chloride transport and the resulting corrosion of steel bars in alkali activated slag concretes," *Materials and Structures*, 49:3663-3677.

[58] Angst, U. M. and Elsener, B. 2017. "The size effect in corrosion greatly influences the predicted life span of concrete infrastructures," *Science Advances*, Vol.3 No.e1700751.

[59] Obla, K. H. Lobo, C. L. and Kim, H. 2016, "Tests and Criteria for Concrete Resistant to Chloride Ion Penetration," *ACI Materials Journal*, 621-631.

[60] Azad, A. K., Ahmad, S. and Azher, S. A. 2007. "Residual Strength of Corrosion- Damaged Reinforced Concrete Beams," *ACI Materials Journal*, Vol. 104 (1): 40-47.

[61] Wang, D., Zhou, X., Fu, B. and Zhang, L.2018. "Chloride ion penetration resistance of concrete containing fly ash and silica fume against combined freezing-thawing and chloride attack," *Construction and Building Materials*, 169: 740-747.

[62] Yuan W., Guo A., Li H., 2017. "Experimental investigation on the cyclic behaviors of corroded coastal bridge piers with transfer of plastic hinge due to non-uniform corrosion," *Soil Dynamics and Earthquake Engineering*, 102:112-123.

140   *Ch. Apostolopoulos and Arg. Drakakaki*

[63] Apostolopoulos C. A., Diamantogiannis G., 2012. "Structural Integrity Problems in Dual-Phase High Ductility Steel Bar," *Applied Mechanical Engineering*, 2(115).

[64] Zhang L., Thomas B., 2003. "Inclusions in continuous casting of steel," *XXIV National Steelmaking Symposium* (Morelia, Mich, Mexico), 138-183.

[65] Kiessling R., 2001 "Nonmetallic Inclusions and their Effects on the Properties of Ferrous Alloys," *Encyclopedia of Materials: Science and Technology*, 6278-6283.

[66] Temmel C., Ingesten N. G., Karlsson B. 2006. "Fatigue anisotropy in cross-rolled, hardened medium carbon steel resulting from MnS inclusions, *Metallurgical and Materials Transactions A*, 37 (10): 2995-3007.

[67] Liu Z., Kobayashi Y., Kuwabara M., Nagai K. 2007. "Interaction between Phosphorous Micro-Segregation and Sulfide Precipitation in Rapidly Solidified Steel- Utilization of Impurity Elements in Scrap Steel, *Materials Transactions*, 48 (1):3079-3087.

[68] Apostolopoulos Alk, Drakakaki A., Konstantopoulos G., Matikas T.2015. "Mapping Sulfides and Strength Properties of Bst420 and B500 Steel Bars Before and After Corrosion," *Humanities & Science University Journal*, 15: 22-32.

[69] Negheimish A. A., Alhozaimy A., Hussain R. R., Al-Zaid R., Singh J. K., 2014. "Role of Manganese Sulfide Inclusions in Steel Rebar in the formation and Breakdown of PaD.D.N. Singh," *NACE International*, 70 (1):74-86.

[70]. Glass G. K., Buenfield N. R. 2000 "Chloride- induced corrosion of steel in concrete," *Progress in Structural Engineering and Materials banner*, 2 (4): 448-458.

[71] Apostolopoulos C. A., Diamantogiannis G., Apostolopoulos A. C., 2015 "Assessment of the mechanical Behavior in Dual-Phase Steel B400c, B450c and B500b in a marine environment," *Journal of Materials in Civil Engineering*, 28 (2).

[72] Li X., Zhao Y. S., Ly H. L., 2015. "Low- cycle fatigue behaviour of corroded and CFRP-wrapped reinforced concrete columns," *Construction and Building Materials*, 101 (1):902-917.

[73] Meda A., Mostosi S., Rinaldi Z., Riva P., 2014. "Experimental evaluation of the corrosion influence on the cyclic behaviour of RC columns, *Engineering Structures*, 76:112-123.

[74] Apostolopoulos Ch., Drakakaki Arg., Basdeki M. 2019. "Seismic Assessment of a Reinforced Concrete Column under Seismic Loads," *International Journal of Structural Integrity*, 10(1): 41-54.

In: Advances in Engineering Research
Editor: Victoria M. Petrova

ISBN: 978-1-53616-092-5
© 2019 Nova Science Publishers, Inc.

Chapter 4

DEVELOPMENT OF AN ONLINE INSPECTION SYSTEM FOR COMPUTER NUMERICAL CONTROL CUTTING LATHE TOOL INSERTS USING EYE-IN-HAND MACHINE VISION

*Wei-Heng Sun[1] and Syh-Shiuh Yeh[2],**
[1]Institute of Mechatronic Engineering,
National Taipei University of Technology, Taipei, Taiwan, R.O.C.
[2]Department of Mechanical Engineering,
National Taipei University of Technology, Taipei, Taiwan, R.O.C.

ABSTRACT

To address the online insert inspection need during computer numerical control (CNC) lathe cutting processes, we developed a system based on eye-in-hand machine vision that incorporates a manipulator. This system can perform online inspections for detection and classification of fractures, built-up edges (BUEs), chipping, and flank wear in the inserts of external turning tools. To circumvent the influence of the machining environment on the acquisition of insert images, a machine vision module equipped with a surrounding light source and a

---

* Corresponding Author's Email: ssyeh@ntut.edu.tw.

fill-light for insert-tips was designed, which could be installed on a manipulator. This manipulator allows the machine vision module to reach the lathe turret position and capture clear and detailed images of the lathe's inserts. Using this strategy, subsequent image processing procedures and insert status judgment are significantly more straightforward. In the process of capturing insert images, the intensity of the surrounding light source and insert-tip fill-light varied so that the surface features and tip features in the insert images could be readily identified. Insert inspection and classification were performed according to the following procedure: 1) construction of the insert's outer profile, 2) capture of the insert's status region, and 3) determination and calculation of the insert's wear region. To obtain images of the insert's wear region, insert images were trimmed according to the vertical flank line, horizontal blade line, and vertical blade line of the insert, which were, in turn, constructed from its outer profile features. The occurrence of flank or chipping wear was determined according to gray-level histograms. The images of the wear region were used to calculate the amount of wear for evaluating and assessing whether an insert had normal or excessive levels of wear. In the present study, insert inspection experiments were performed to establish that the developed machine vision module and computational methods could detect insert fractures, BUEs, chipping, and flank wear. When the positioning offset of the manipulator-operated machine vision module was lower than 0.5 mm, the chipping rate detection error was less than 8%, and the detection error of the amount of wear was within 2 pixels. These results establish the viability and efficacy of the developed online eye-in-hand lathe inspection system.

Keywords: online inspection, tool inserts, CNC lathe machines, eye-in-hand, machine vision

INTRODUCTION

In many conventional industrial lathe insert applications, the degree of lathe insert wear is often assessed after degradation of product quality. To improve product quality, numerous modern factories incorporate a variety of autonomous devices to monitor production data. The goal of these devices is to either detect defective products using a variety of sensors or identify the likely causes of defective products using data analysis, assisting in addressing the problems prior to production [1]. This study

Development of an Online Inspection System for Computer ... 145

aims to classify common types of wear using eye-in-hand machine vision and expand the scope of this method via manipulator operation. This will allow detection and evaluation of the insert condition during its application to satisfy the requirements of actual manufacturing processes. Generally, the wear mechanism of lathe inserts includes abrasion wear, diffusion wear, oxidative wear, fatigue wear, and adhesive wear. The four most common wear scenarios in cutting processes are flank wear, chipping, fractures, and built-up edges (BUEs). Most of these wears are manifested at the tip and flank of a tool [2–5]. In this study, the classification of insert conditions is based on these scenarios. In addition, eye-in-hand visual inspections were used to inspect the state of wear of the inserts.

The inspection methods used in previous studies for the detection of the different forms of insert wear may be categorized into two: indirect inspections based on the analysis of lathe-related data transmitted from external sensors [6, 7] and direct inspections based on actual measurements of a cutting tool's condition [8, 9]. In indirect inspections, the state of a cutting tool is determined via data analysis. In most reports on indirect inspection methods, analyses are performed on a few lathe states, and the cutting status is determined using systematic approaches. The aim of these methods is to obviate the subjective judgments from experienced operators, which assists in minimizing human errors and increasing the capacity for automated production [10]. For example, the state of tool wear could be analyzed using observed changes in cutting noise or vibrations [11, 12]. Moreover, cutting tool monitoring could be performed by measuring the cutting temperature and forces [13, 14], and the cutting status might be analyzed by varying the power or current signals from a lathe [15]. All of these inspection methods are based on the detection and analysis of sensing signals. In recent years, the widespread development of charge-coupled device (CCD) cameras has resulted in the development of indirect inspection methods based on the captured image, in which the surface texture of a workpiece is analyzed to determine the status of a cutting tool. Thus, changes in the surface roughness and surface texture of a workpiece are used to determine the occurrence of wear [16–20]. In several publications, more advanced forms of monitoring systems have been developed based on the fusion of multiple sensors and visual information from images [21, 22]. Moreover, different algorithms are used to improve the accuracy of the monitoring and evaluation tool [16, 17, 23, 24]. On the basis of these reports, it may be inferred that the state of the cutting tool can be determined by analyzing the changes in the lathe's signals. However, indirect inspection systems are susceptible to the external sensing environment that consequently reduces their accuracy [25]. Nevertheless, direct inspection methods for detecting changes in the status of cutting tools have been developed to circumvent this issue.

In most direct inspection methods, the problems associated with the cutting processes are analyzed via direct observations of the actual status of the cutting tools. Current literature reveals that several models of cutting tools have been established using sound and light or probes used to observe the status of the cutting tools [21, 26, 27]. However, it is challenging to apply these methods to on-site inspections because of the complexity of the required measuring equipment. An alternative method is to analyze the tool status based on images captured with a CCD camera. These tools may be categorized into two: the first category is the wear detection and analysis based on the outer contour and profile of the tools [28], in which usability is determined by monitoring the state of wear on the outer profile of the tools. The second category is based on surface texture or surface roughness analyses [29]. This method yields highly detailed information on cutting processes, and it is used to perform finely differentiated inspections and evaluations of the cutting tools and lathes. Many CCD-camera-based visual inspections are used to analyze different insert locations. A few previous studies have focused on the crater wear analysis [30, 31], focusing on other CCD-camera-based investigations for monitoring and analysis of flank wear [32]. In most cases, visual wear detection can be used to obtain different types of cutting tool information of images to conveniently detect changes in the outer profile and surface texture. Several researchers (including Dutta et al. [17]) have proposed different methods of detecting cutting tool wear, like those used by Giusti et al. [33]. Moreover, Rangwala and Dornfeld [34] proposed the detection of cutting tool wear through visual methods. In their studies, neural networks were used to analyze the state of wear. To optimize the extraction of wear features, Kurada and Bradley [35] proposed a method wherein gradient operators were used to identify texture features. In this method, an octagonal matrix and a slope (constructed from the brightness differences and radial distances from the center of the matrix) are used to search for features on the boundary of the wear region, which are used to obtain the optimized wear feature locations. In this procedure, image smoothing is performed during preprocessing to reduce the irregular interference. Then, the features are identified using image thresholding to convert the pixel values into a binary image, which assist in providing information on the actual degree of wear-associated changes. Yuan et al. [36] proposed a new edge detection method based on wavelet transforms and a novel filtering method to obtain the average of an image. The authors also proposed a new wavelet function to describe the gray-level changes of an image and reduce noise-related interference for improving the extraction of edge features. This wavelet function is used to quantify the width, length, and center of abrasion regions. Moreover, Wang et al. [37] proposed an image processing procedure using a rough-to-fine thresholding strategy different from conventional constant thresholding-based methods. First, binary images are obtained to search for candidate edge points at the bottom of the wear region, and then, threshold-independent edge detection methods based on moment invariance are used to determine the wear edges. To reduce computational time, a critical area is determined on a preliminary basis to limit the region-of-interest of all of the subsequent processes, thus circumventing the threshold-dependent methods when detecting the wear features. Li et al. [38] investigated the pulse-coupled neural networks, derived from the field of bionics and used to monitor the cutting tool wear. On the basis of the assumption that the gray intensity of the field-based tool wear is higher than that of the body of the tool and its background, binary images are segmented using similar and spatially adjacent gray pixel clusters. This assists in detecting the field-based tool wear and evaluating the cutting tool state of wear. Shahabi and Ratnam [39] established a wear inspection method that did not require precise tool alignment of the cutting tools. In this approach, the external profile of the original images is aligned with the images of the tool, and an algorithm based on Wiener filtering, median filtering, morphological operations, and thresholding is applied to reduce the system errors caused by cutting tool misalignment, the presence of micro-dust particles, ambient light vibrations, and intensity variations. With this technique, it is possible to obtain the tool holder position and its positioning errors. Pfeifers and Wiegers [40] exploited variations of the intensity of the light source to search for wear edge features and address issues related to the tool-positioning problem and the adhesion of the contaminants of the cutting tools due to machining processes. Changes in lighting induce corresponding changes in the shadows of the cutting tool's wear edges. Although the actual position of the wear edge may remain unchanged, a high-pass filter and image thresholding yield information on cutting tool wear under different illumination sources. Moreover, overlap operations can be used to extract the edge positions that appear repeatedly to determine the positions of their strong edges. This procedure filters contaminant-induced errors and reduces the effects and the influence of shadowing changes to the inspection system.

Unlike the methods proposed in previous studies, the eye-in-hand online insert inspection system based on the on-machine insert condition monitoring system developed by the authors [41] can be used to analyze flank wear. In addition, contour and texture detection methods were fused to develop a highly accurate evaluation and examination system. This system is suitable for lathe insert online inspection and prevention of several environmental issues commonly encountered during these inspections. As part of this investigation, a machine vision module for manipulation has been developed, which permits the module to reach the turret position inside CNC lathes for insert inspection. This also increases the flexibility of the proposed system during online lathe insert inspections. The machine vision module includes a CCD camera and lens for capturing insert images. It also includes a protective box to protect the photographic equipment and peripheral circuitry from the scrap metal splatter and prevent the lens from being contaminated with cutting fluids. The module has a cleaning air tube that can be used to remove some of the surface contaminants on an insert. This alleviates a few of the difficulties in the

Development of an Online Inspection System for Computer ... 149

subsequent image processing and increases the accuracy of insert condition assessments. The developed module also includes a surrounding light source and a fill-light for insert-tips, which allows the state of an insert to be examined by varying the intensity of two light sources. Using this system, the location of the insert's blade and tip can be determined by modulating the light sources. This approach can be used during insert state inspections with lower levels of alignment precision between the tool and manipulator, thus improving the viability of the developed system for image recognition inside lathes. Insert images were also captured at different levels of light source intensities to reduce the influence of the external environment on the results of the insert status inspections. This allows the system to obtain accurate inspection results at different environmental conditions. It also reduces the insufficient lighting or solarization caused by differences in the insert condition.

In this study, captured insert images that contained wear features were analyzed and detected. In addition, common insert wear states (fractures, BUEs, chipping, and flank wear) based on the distribution of wear textures and features of these images were quantified. Using this system, experiments were performed with inserts under different conditions. It was demonstrated that the developed system can be used to perform accurate and stable insert status inspections.

PROCEDURE FOR INSPECTION AND CLASSIFICATION OF INSERTS

Procedure to Capture Insert Images

The image capture is performed by applying fill-light on the inspected insert, and the intensity of the fill-light varies according to the appropriate feature insert strength regardless of its type and condition. Fill-light was applied using two sets of light sources at different positions: a surrounding light source and a fill-light for the tip of the inserts. The main purpose of the surrounding light source is to capture the surface morphology and area features during the insert inspections. The insert-tip fill-light is used to enhance the tip status features for subsequent post-processing procedures and analyses. In the developed image capturing procedure, the high-intensity surrounding light source is first used to photograph the insert and capture exposure images of the flank's profile. The surrounding light source and insert-tip fill-light are used in tandem to photograph the insert and capture the exposure images of the insert's profile. After the exposure images have been captured, this study starts by capturing the feature images. The insert-tip fill-light is first turned off, and the intensity of the surrounding light source is adjusted to obtain the flank's feature image with proper levels of exposure. Then, the intensity of the insert-tip fill-light is adjusted to obtain properly exposed images. This study begins analyzing the inserts after the exposure images and feature images have been captured and stored.

Construction of Inserts' Outer Profile

First, the flank profile features were determined by performing grayscale image thresholding on the flank exposure image. Similarly, the profile features of the insert were determined via grayscale image thresholding of the insert exposure image. Then, Hough transform was used to determine the straight lines of the obtained binary images. The vertical flank line and horizontal blade line were determined from the binary exposure image of the flank's profile, whereas the vertical blade line was determined from the binary exposure image of the insert's profile. The horizontal blade line and the vertical blade line can be used to trim and rotate the binary images, thus forming a complete binary image of the insert's outer profile. The vertical flank line was used to segment a complete binary image of the insert's outer profile into two zones: Zone B, which comprised the area below the front edge of the tip, and Zone A, which was the area below the rear edge of the insert. These zones will be used for the lathe insert feature recognition.

Development of an Online Inspection System for Computer ... 151

Capture of Insert's Status Region

The horizontal blade line can be used to determine whether either or both fracture and BUE have occurred in an insert. Gray image thresholding of the trimmed image of the insert's outer profile yields a binary insert image, and segmenting this image using the horizontal blade line returns the fracture zone and BUE zone. The sizes of the pixel areas in the fracture zone and BUE zone can be used to assess whether a lathe insert is fractured or has a BUE.

If the lathe inserts inspected by the developed system are neither fractured nor have a BUE, this study determines whether flank wear has occurred. The grayscale conversion is performed on the outer trimmed profile image of the insert. In this study, the averaged image RGB values are used in grayscale processing. After the insert's outer profile image is converted to a grayscale image, edge detection was performed using the Sobel operator to effectively extract the edge features. Zone A was removed from the zonally divided image of the insert's outer profile to isolate the area containing the flank wear features. To trim the flank wear sections of the insert for subsequent assessment and calculations, noise removal computations, contrast stretching processes, and erosion and dilation operations were performed on the trimmed outer profile image of the insert, assisting in obtaining the image corresponding to the area of the flank wear zone. Finally, the binary image of the flank wear zone was used to perform trimming operations on the image of the insert's outer profile to obtain a photograph of the flank wear zone.

Determination and Calculation of the Wear Region of Inserts

The trimmed image of the lathe insert's wear region can be used to determine whether an insert has flank wear or chipping. Flank wear is associated with a tear because of rubbing between a workpiece and the insert during machining processes, consequently making the surface features of flank wear continuous and smooth. Meanwhile, chipping is associated with tip breakages induced by abnormalities in the machining process; therefore, the surfaces associated with chipping are more rugged. Hence, this study analyzed the continuity of the surface features in the photographs of the wear region to determine whether the wear region of the insert exhibits flank wear or chipping. When the image of the flank wear zone is converted into a grayscale image, the gray-level histogram of the pixel points is obtained. The presence of flank wear or chipping was assessed according to the percentage of pixels that exceeded a defined threshold value among the pixels that represented the wear region. In addition, the pixel length between the upper and lower boundaries was used to calculate the amount of wear, and the pixel units were converted to a physical unit of length (mm) according to the shooting angle of the wear region. In Figure 1, as the length of each pixel in the wear region image is converted to 0.007 mm, the upper and lower boundaries of the wear region are separated by 184 pixels, and the amount of wear shown is 1.288 mm.

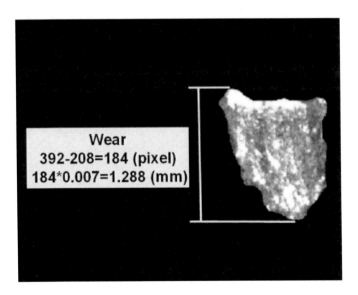

Figure 1. Results of the calculation of the amount of wear based on a trimmed photograph of the wear region [41].

Development of an Online Inspection System for Computer ... 153

EXPERIMENT ON EYE-IN-HAND ONLINE INSERT INSPECTION SYSTEM

This study has designed an adjustable and movable method and devices for lathe insert inspection to aid factory automation. The method assists in developing CNC lathe insert inspections for the flexibility of factory automation. The machine vision module is operated using a six-degree-of-freedom (6-DOF) manipulator, commonly used in production lines, by integrating the machine vision module into the manipulator. We also designed an eye-in-hand online insert inspection system that could be used to inspect lathe inserts in various positions. Figure 2 illustrates the relative positions of the manipulator and CNC lathe during online inspections. The posture of the manipulator is adjusted carefully when the machine vision module of the eye-in-hand online insert inspection system is moved to prevent collisions with the internal components of the CNC lathe. The design of the machine vision module is based on the spatial restrictions of the lathe and the contamination feasibility caused by the machining environment, as shown in Figure 2. The machine vision module includes a protective box, which encloses the photographic hardware to protect the equipment and prevent lens contamination from the internal machining environment. A cleaning air tube was used to clean the inserts to capture clear images of the insert during the inspection processes. The protective box was equipped with an adjustable LED surrounding light source, sealed with epoxy resin. A connecting fixture was set up at the bottom of the protective box, and it was used to fix the protective box and the machine vision module to the end frame of the manipulator. The machine vision module used a GigE DFK 23 GP031 color industrial camera and a visual resolution of 2592×1944 (15 fps). The module uses a Myutron HS3514J CCTV lens with a doubler lens to ensure that clear images of the lathe insert's features are captured. A $90°$ reflecting mirror permits adjustments of the shooting angle of the camera.

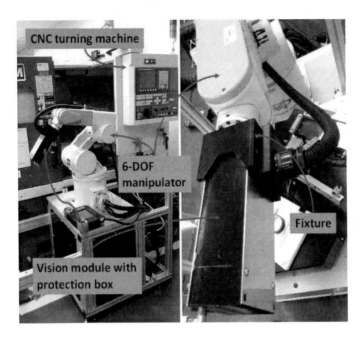

Figure 2. The eye-in-hand online insert inspection system formed by combining a manipulator with a machine vision module.

To establish the viability of the developed online insert inspection system, this study first installed the machine vision module inside a CNC lathe; then, an experiment was performed using 20 inserts under varying conditions. The experimental results indicate that the developed system accurately determines the conditions of each insert [41]. To improve the flexibility of the developed system for the classification and inspection of lathe inserts, this study implements the developed machine vision module and insert inspection method in a 6-DOF manipulator to construct an eye-in-hand online insert inspection system. An experiment was performed to validate the usability of the developed inspection system, shown in Figure 3. A variety of positions for the machine vision module were tested in the eye-in-hand online insert inspection system to determine the impact of the module's position on the insert inspections. More specifically, the effects of the insert alignment position were tested by applying positioning offsets to the optimal position that were either "small" (0.25 mm) or "large" (0.5 mm). Sixteen insert alignment positions were measured, and comparative

Development of an Online Inspection System for Computer ... 155

analyses were performed on the results. In addition, a systematic validation was conducted by performing these measurements on two insert wear states, as shown in Table 1 and Table 2. On the basis of the data in Table 1 and Figure 4, standard deviations of 1.410 and 0.661 were observed in the chipping rate and amount of wear, respectively, based on the offset of 0.25 mm. When the offset was 0.5 mm, the standard deviations of the chipping rate and amount of wear were 2.348 and 0.696, respectively. Table 2 and Figure 5 indicated that the standard deviation of the chipping rate was 0.663 and the amount of wear of the chipping rate was 1.497 when the offset was 0.25 mm, whereas the standard deviations of the chipping rate and amount of wear were 1.728 and 0.800, respectively, when the offset was 0.50 mm. In Set II (chipping wear), the complete chipping features were not captured in the insert alignment positions; the data corresponding to these measurements were omitted from the calculations, and this was represented by (.) in Table 2.

These data demonstrate that positioning offsets induce noticeable changes in the measurements. This is caused by differences in the lighting angle of the inserts in each position, which produce marginal changes in the resulting photographs that subsequently affect the final results of the inspections. In particular, "large" (0.50 mm) offsets induced the most significant differences. The experimental results indicate that 0.50 mm offsets generally induce larger differences than 0.25 mm offsets. This is because chipping rate calculations are performed by analyzing the gray-level histograms of flank wear zone images. Although the algorithm and the operation of the light sources are identical regardless of the offset, the instant at which an image is captured is susceptible to alterations in lighting. Hence, these offsets induce minor alterations in the gray-level histograms of flank wear zone images. Nonetheless, as long as the offsets were within 0.25 mm, chipping rate errors did not exceed 3%. Even when the larger chipping errors were induced by a 0.50 mm offset, the errors in chipping rate detection were within 8%. The detection errors for the wear always remained 2 pixels regardless of the offset. Hence, the eye-in-hand online insert inspection system is beneficial to a certain degree of positioning offset in the machine vision module. In addition, this experiment demonstrates that the positioning offset of the machine vision module is maintained within 0.50 mm to obtain an acceptable quality measurement. It was also demonstrated that the developed system offered a certain degree of variability in the insert alignment position of the machine vision module. Therefore, the eye-in-hand online insert inspection system can be used to perform lathe insert inspections without highly precise alignment and positioning in the machine vision module. This study has established the robustness of the developed system against changes in the environment or insert. The developed system is applicable for online insert inspections in CNC lathes.

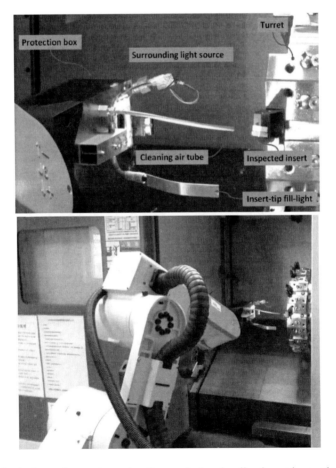

Figure 3. Lathe insert inspection using the eye-in-hand online insert inspection system.

Development of an Online Inspection System for Computer ... 157

Table 1. Results of the positioning offset experiment (Set I)

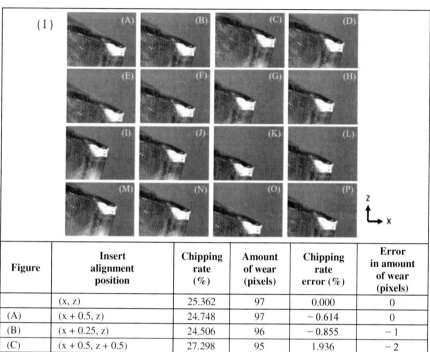

Figure	Insert alignment position	Chipping rate (%)	Amount of wear (pixels)	Chipping rate error (%)	Error in amount of wear (pixels)
	(x, z)	25.362	97	0.000	0
(A)	(x + 0.5, z)	24.748	97	− 0.614	0
(B)	(x + 0.25, z)	24.506	96	− 0.855	− 1
(C)	(x + 0.5, z + 0.5)	27.298	95	1.936	− 2
(D)	(x + 0.25, z + 0.25)	24.180	95	− 1.181	− 2
(E)	(x + 0.5, z − 0.5)	28.661	96	3.299	− 1
(F)	(x + 0.25, z − 0.25)	27.021	97	1.658	0
(G)	(x − 0.5, z)	27.550	95	2.187	− 2
(H)	(x − 0.25, z)	26.958	96	1.590	− 1
(I)	(x − 0.5, z + 0.5)	31.437	96	6.075	− 1
(J)	(x − 0.25, z + 0.25)	26.239	95	0.876	− 2
(K)	(x − 0.5, z − 0.5)	32.591	95	7.229	− 2
(L)	(x − 0.25, z − 0.25)	28.134	96	2.771	− 1
(M)	(x, z + 0.5)	27.700	96	2.338	− 1
(N)	(x, z + 0.25)	25.144	95	−0.218	− 2
(O)	(x, z − 0.5)	29.950	95	4.587	− 2
(P)	(x, z − 0.25)	27.907	96	2.544	− 1
Average and maximum absolute error		Average value of (A)–(P)		Maximum absolute error in (A)–(P)	
		27.501	95.687	7.229	2
Standard deviation		2.300	0.681	2.299	0.681

Table 2. Results of the positioning offset experiment (Set II)

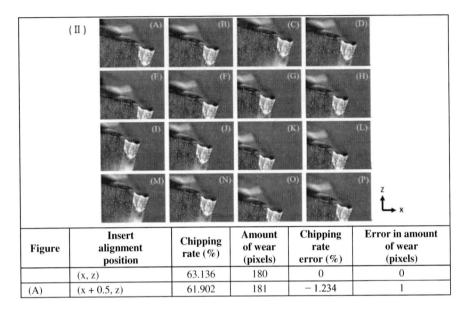

Figure	Insert alignment position	Chipping rate (%)	Amount of wear (pixels)	Chipping rate error (%)	Error in amount of wear (pixels)
	(x, z)	63.136	180	0	0
(A)	(x + 0.5, z)	61.902	181	− 1.234	1

Figure	Insert alignment position	Chipping rate (%)	Amount of wear (pixels)	Chipping rate error (%)	Error in amount of wear (pixels)
(B)	(x + 0.25, z)	63.555	178	0.419	− 2
(C)	(x + 0.5, z + 0.5)	62.975	180	− 0.16	0
(D)	(x + 0.25, z + 0.25)	63.254	180	0.118	0
(E)	(x + 0.5, z − 0.5)	(57.495)	(116)	(− 5.641)	(− 64)
(F)	(x + 0.25, z − 0.25)	(63.583)	(153)	(− 0.447)	(− 27)
(G)	(x − 0.5, z)	65.919	182	2.783	2
(H)	(x − 0.25, z)	64.159	182	1.023	2
(I)	(x − 0.5, z + 0.5)	66.382	180	3.246	0
(J)	(x − 0.25, z + 0.25)	63.572	178	0.436	− 2
(K)	(x − 0.5, z − 0.5)	(64.693)	(121)	(1.557)	(− 59)
(L)	(x − 0.25, z − 0.25)	(63.298)	(166)	(0.162)	(− 14)
(M)	(x, z + 0.5)	65.055	180	1.919	0
(N)	(x, z + 0.25)	62.149	180	− 0.986	0
(O)	(x, z − 0.5)	(61.619)	(133)	(− 1.516)	(− 47)
(P)	(x, z − 0.25)	(63.851)	(163)	(0.715)	(− 17)
Average and maximum absolute error		Average value of (A)–(P)		Maximum absolute error in (A)–(P)	
		63.892	180.1	7.229	2
Standard deviation		1.421	1.3	1.4208	1.3

Development of an Online Inspection System for Computer ... 159

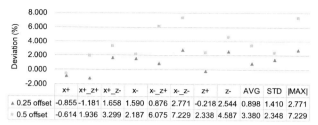

(a) Comparison between chipping rate errors associated with positioning offsets of 0.25 and 0.5 mm

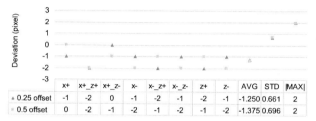

(b) Comparison between errors in the amount of wear associated with positioning offsets of 0.25 and 0.5 mm

Figure 4. Comparison between errors in chipping rate and amount of wear with different positioning offsets (corresponding to Table 1).

(a) Comparison between chipping rate errors associated with positioning offsets of 0.25 and 0.5 mm

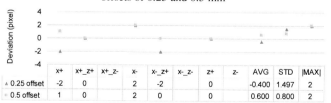

(b) Comparison between errors in the amount of wear associated with positioning offsets of 0.25 and 0.5 mm

Figure 5. Comparison between errors in chipping rate and amount of wear with different positioning offsets (corresponding to Table 2).

CONCLUSION

In the CNC lathe machining processes, lathe inserts are subject to fractures, BUEs, chipping, or flank wear owing to the operation of the lathe and the cutting conditions. Previous insert inspection methods tend to involve a multitude of procedures to reduce the productivity of machining processes, increase production costs, and reduce cutting precision. This study has addressed these problems by combining a manipulator with a machine vision module to create an eye-in-hand online lathe inspection system. This system is capable of detecting the four aforementioned forms of insert wear and calculating the amount of wear to determine whether an insert exhibits normal wear or excessive wear.

To allow the developed inspection system to reach the turret position of a CNC lathe through a manipulator operation for online insert inspection, the developed machine vision module includes a protective box, a cleaning air tube, and two types of light sources: a surrounding light source and a fill-light for insert-tips. The protective box protects the lens from contamination with splattered metal shavings and cutting fluids, and the cleaning air tube can be used to prepare an insert inspection by removing some of the surface contaminants. The variable-intensity surrounding light source and insert-tip fill-light can be used to analyze the effects of the variations in the light source based on visual insert inspections. During insert image capture, the intensities of the surrounding light source and insert-tip fill-light were adjusted to ensure that the inspected insert exhibited an adequate level of feature strength. More specifically, the surrounding light source was used to illuminate the surface of the insert with highly intense illumination to capture its surface morphology and area features. The insert-tip fill-light was used to enhance the status features of the inserts' tips for subsequent post-processing procedures and analyses.

The insert inspection and classification procedure designed in this study is as follows: 1) construction of the insert's outer profile, 2) capture of the insert's status region, and 3) determination and calculation of the wear region of the insert. To construct the insert's outer profile, the outer

Development of an Online Inspection System for Computer ... 161

profile features were determined from the exposure images; subsequently, a vertical flank line, horizontal blade line, and vertical blade line were constructed using these features. The insert image was then trimmed according to these lines to prepare the image for subsequent insert-feature-recognition processes. The capture of an insert's status region was performed to determine the insert's fracture and BUE zones according to its outer profile feature lines. The outer profile image of the insert was trimmed to obtain a photograph of the insert's wear region. The determination and calculation of the insert's wear region were performed to ascertain the presence of flank wear or chipping wear in the trimmed photograph of the wear region based on the gray-level histogram of the pixels in the photograph. The pixel length between the upper and lower boundaries of the wear region was used to calculate the amount of wear, serving as an indicator for determining the state of wear (normal or excessive wear).

The developed machine vision module and insert inspection methods were combined with a manipulator to construct the eye-in-hand online insert inspection system, and an online lathe insert inspection experiment was conducted using a CNC lathe. The results of this experiment demonstrated that the developed machine vision module and insert inspection method could be used to detect the presence of fractures, BUEs, chipping, and flank wear in lathe inserts. In addition, the errors in the detection of the chipping rate were < 3% and < 8% with positioning offsets of 0.25 and 0.5 mm, respectively, in the machine vision module. The errors detected in the wear were within 2 pixels regardless of the offset. This result establishes the viability and operational performance of the proposed software and hardware system.

ACKNOWLEDGMENTS

This research was funded in part by the Ministry of Science and Technology, Taiwan, R.O.C., under Contract MOST 104-2221-E-027-132 and MOST 103-2218-E-009-027-MY2. The authors would like to thank representatives from the SRAM Taiwan Company for their useful discussions with the research team. The authors especially thank to Meng-Hui Lin (SRAM Taiwan Company) for his beneficial discussions.

REFERENCES

[1] Jones, B. E. "Sensors in Industrial Metrology." *Journal of Physics E: Scientific Instruments*, 20, no. 9, (1987), 1113-16.

[2] Avinash, C.; Raguraman, S.; Ramaswamy, S.; Muthukrishnan, N. "An Investigation on Effect of Workpiece Reinforcement Percentage on Tool Wear in Cutting Al-SiC Metal Matrix Composites." Paper presented at the *ASME International Mechanical Engineering Congress and Exposition, Proceedings*, 2008.

[3] Akbar, F.; Mativenga, P. T.; Sheikh, M. A. "An Evaluation of Heat Partition in the High-Speed Turning of AISI/SAE 4140 Steel with Uncoated and TiN-Coated Tools." *Proceedings of the Institution of Mechanical Engineers, Part B: Journal of Engineering Manufacture*, 222, no. 7, (2008), 759-71.

[4] Ee, K. C.; Balaji, A. K.; Jawahir, I. S. "Progressive Tool-Wear Mechanisms and Their Effects on Chip-Curl/Chip-Form in Machining with Grooved Tools: An Extended Application of the Equivalent Toolface (ET) Model." *Wear*, 255, no. 7-12, (2003), 1404-13.

[5] Nordgren, A.; Melander, A. "Tool Wear and Inclusion Behaviour During Turning of a Calcium-Treated Quenched and Tempered Steel Using Coated Cemented Carbide Tools." *Wear*, 139, no. 2, (1990), 209-23.

[6] Massol, O; Li, X.; Gouriveau, R.; Zhou, J. H.; Gan, O. P. "An ExTS Based Neuro-Fuzzy Algorithm for Prognostics and Tool Condition Monitoring." Paper presented at the 11th International Conference on Control, Automation, Robotics and Vision, *ICARCV*, 2010, 2010.

Development of an Online Inspection System for Computer ... 163

[7] Ralston, P. A. S.; Ward, T. L.; Stottman, D. J. C. "Computer Observer for in-Process Measurement of Lathe Tool Wear." *Computers and Industrial Engineering*, 15, no. 1-4, (1988), 217-22.

[8] Rutelli, G; Cuppini, D. "Development of Wear Sensor for Tool Management System." *Journal of Engineering Materials and Technology, Transactions of the ASME*, 110, no. 1, (1988), 59-62.

[9] Novak, A; Wiklund, H. "Reliability Improvement of Tool-Wear Monitoring." *CIRP Annals - Manufacturing Technology*, 42, no. 1, (1993), 63-66.

[10] Szélig, K; Alpek, F; Berkes, O; Nagy, Z. "Automatic Inspection in a CIM System." *Computers in Industry*, 17, no. 2-3, (1991), 159-67.

[11] Downey, J.; Bombiński, S.; Nejman, M.; Jemielniak, K. "Automatic Multiple Sensor Data Acquisition System in a Real-Time Production Environment." *Procedia CIRP*, 33, (2015), 215-220.

[12] Scheffer, C.; Heyns, P. S. "Monitoring of Turning Tool Wear Using Vibration Measurements and Neural Network Classification." Paper presented at the *25th International Conference on Noise and Vibration Engineering, ISMA*, 2000.

[13] Prasad, B. S.; Prabha, K. A.; Kumar, P. V. S. G. "Condition Monitoring of Turning Process Using Infrared Thermography Technique – an Experimental Approach." *Infrared Physics and Technology*, 81, (2017), 137-47.

[14] Fu, P.; Li, W.; Guo, L. "Fuzzy Clustering and Visualization Analysis of Tool Wear Status Recognition." *Procedia Engineering*, 23, (2011), 479-486.

[15] Hamade, R. F.; Ammouri, A. H. "Current Rise Index (CRI) Maps of Machine Tool Motors for Tool-Wear Prognostic." Paper presented at the *ASME 2011 International Mechanical Engineering Congress and Exposition, IMECE*, 2011, 2011.

[16] Dutta, S; Pal, S. K.; Sen, R. "Progressive Tool Flank Wear Monitoring by Applying Discrete Wavelet Transform on Turned Surface Images." *Measurement: Journal of the International Measurement Confederation*, 77, (2016), 388-401.

[17] Dutta, S; Pal, SK; Mukhopadhyay, S; Sen, R. "Application of Digital Image Processing in Tool Condition Monitoring: A Review." *CIRP Journal of Manufacturing Science and Technology*, 6, no. 3, (2013), 212-32.

[18] Dutta, S; Datta, A; Chakladar, ND; Pal, SK: Mukhopadhyay, S; Sen, R. "Detection of Tool Condition from the Turned Surface Images Using an Accurate Grey Level Co-Occurrence Technique." *Precision Engineering*, 36, no. 3, (2012), 458-66.

[19] Kwon, Y.; Ertekin, Y.; Tseng, T. L. "Characterization of Tool Wear Measurement with Relation to the Surface Roughness in Turning." *Machining Science and Technology*, 8, no. 1, (2004), 39-51.

[20] Kassim, A. A.; Mannan, M. A.; Jing, Ma. "Machine Tool Condition Monitoring Using Workpiece Surface Texture Analysis." *Machine Vision and Applications*, 11, no. 5, (2000), 257-63.

[21] Prasad, B. S.; Sarcar, M. M. M.; Ben, B. S. "Surface Textural Analysis Using Acousto Optic Emission- and Vision-Based 3D Surface Topography-a Base for Online Tool Condition Monitoring in Face Turning." *International Journal of Advanced Manufacturing Technology*, 55, no. 9-12, (2011), 1025-35.

[22] Prasad, B. S.; Sarcar, M. M. M. "Experimental Investigation to Predict the Condition of Cutting Tool by Surface Texture Analysis of Images of Machined Surfaces Based on Amplitude Parameters." *International Journal of Machining and Machinability of Materials*, 4, no. 2-3, (2008), 217-36.

[23] Bhat, N. N.; Dutta, S.; Pal, S. K.; Pal, S. "Tool Condition Classification in Turning Process Using Hidden Markov Model Based on Texture Analysis of Machined Surface Images." *Measurement: Journal of the International Measurement Confederation*, 90, (2016), 500-09.

[24] Mannan, M. A.; Mian, Z.; Kassim, A. A. "Tool Wear Monitoring Using a Fast Hough Transform of Images of Machined Surfaces." *Machine Vision and Applications*, 15, no. 3, (2004), 156-63.

Development of an Online Inspection System for Computer ...   165

[25] Bartow, M. J.; Calvert, S. G.; Bayly, P. V. "Fiber Bragg Grating Sensors for Dynamic Machining Applications." Paper presented at the *SPIE - The International Society for Optical Engineering*, 2003.

[26] Wang, W. H.; Wong, Y. S.; Hong, G. S. "3D Measurement of Crater Wear by Phase Shifting Method." *Wear*, 261, no. 2, (2006), 164-71.

[27] Dawson, T. G.; Kurfess, T. R. "Quantification of Tool Wear Using White Light Interferometry and Three-Dimensional Computational Metrology." *International Journal of Machine Tools and Manufacture*, 45, no. 4-5, (2005), 591-96.

[28] Klancnik, S.; Ficko, M.; Balic, J.; Pahole, I. "Computer Vision-Based Approach to End Mill Tool Monitoring." *International Journal of Simulation Modelling*, 14, no. 4, (2015), 571-83.

[29] Kerr, D.; Pengilley, J.; Garwood, R. "Assessment and Visualisation of Machine Tool Wear Using Computer Vision." *International Journal of Advanced Manufacturing Technology*, 28, no. 7-8, (2006), 781-91.

[30] Lanzetta, M. "A New Flexible High-Resolution Vision Sensor for Tool Condition Monitoring." *Journal of Materials Processing Technology*, 119, no. 1-3, (2001), 73-82.

[31] Giusti, F.; Santochi, M.; Tantussi, G. "On-Line Sensing of Flank and Crater Wear of Cutting Tools." *CIRP Annals - Manufacturing Technology*, 36, no. 1, (1987), 41-44.

[32] Bahr, B; Motavalli, S; Arfi, T. "Sensor Fusion for Monitoring Machine Tool Conditions." *International Journal of Computer Integrated Manufacturing*, 10, no. 5, (1997), 314-23.

[33] Giusti, F; Santochi, M; Tantussi, G. "A Flexible Tool Wear Sensor for NC Lathes." *CIRP Annals - Manufacturing Technology*, 33, no. 1, (1984), 229-32.

[34] Rangwala, S.; Dornfeld, D. "Integration of Sensors Via Neural Networks for Detection of Tool Wear States." Paper presented at the American Society of Mechanical Engineers, Production Engineering Division (Publication) PED, 1987.

[35] Kurada, S.; Bradley, C. "A Machine Vision System for Tool Wear Assessment." *Tribology International*, 30, no. 4, (1997), 295-304.

[36] Yuan, Q; Ji, SM; Zhang, L. "Study of Monitoring the Abrasion of Metal Cutting Tools Based on Digital Image Technology." Paper presented at the *SPIE - The International Society for Optical Engineering*, 2005.

[37] Wang, W. H.; Hong, G. .S.; Wong, Y. S. "Flank Wear Measurement by a Threshold Independent Method with Sub-Pixel Accuracy." *International Journal of Machine Tools and Manufacture*, 46, no. 2, (2006), 199-207.

[38] Li, P.; Li, Y.; Yang, M; Zheng, J.; Yuan, Q. "Monitoring Technology Research of Tool Wear Condition Based on Machine Vision." Paper presented at the *World Congress on Intelligent Control and Automation (WCICA)*, 2008.

[39] Shahabi, H. H.; Ratnam, M. M. "On-Line Monitoring of Tool Wear in Turning Operation in the Presence of Tool Misalignment." *International Journal of Advanced Manufacturing Technology*, 38, no. 7-8, (2008), 718-27.

[40] Pfeifer, T.; Wiegers, L. "Reliable Tool Wear Monitoring by Optimized Image and Illumination Control in Machine Vision." *Measurement: Journal of the International Measurement Confederation*, 28, no. 3, (2000), 209-18.

[41] Sun, W. H.; Yeh, S. S. "Using the Machine Vision Method to Develop an on-Machine Insert Condition Monitoring System for Computer Numerical Control Turning Machine Tools." *Materials*, 11, no. 10, (2018).

In: Advances in Engineering Research
Editor: Victoria M. Petrova

ISBN: 978-1-53616-092-5
© 2019 Nova Science Publishers, Inc.

Chapter 5

TYPE-2 FUZZY LOGIC CONTROL FOR QUADROTOR

Gherouat Oussama[1,], Fayssal Amrane[2,*] and Abdelouahab Hassam[1]*

[1]Laboratory of Intelligent Systems (LSI), Department of Electronic, University of Setif 1, Setif, Algeria

[2]Automatic Laboratory of Setif (LAS), Department of Electrical Engineering, University of Setif 1, Setif, Algeria

ABSTRACT

In this chapter Type-2 Fuzzy logic Control (T2-FLC) is proposed in order to control the altitude and attitude of an unmanned quadrotor. The mathematical model of six degree of freedom (6DOF) aerial vehicle is nonlinear and highly-coupling, and under-actuated; all these drawbacks make the system unstable. For the whole-system; the partials six controls were distributed and select the membership's functions carefully due of the impact on each other. However, the memberships functions (MFs) are the heart of any fuzzy logic system the fuzzy type-2 by nature can handle better uncertainty and solve the strong coupling and trajectory tracking

---

[*] Corresponding Author's Emails: gherouat19@gmail.com, abdelhassam@univ-setif.dz; amrane_fayssal@live.fr.

168   *Gherouat Oussama, Fayssal Amrane and Abdelouahab Hassam*

problem. In order to assess the fuzzy type-2 controller performance, it has been compared with a Sliding Mode Controller (SMC) for two popular different trajectories straight lines and Helical. The proposed control based on T2-FLC is used to overcome the drawbacks of the classical control based on SMC in terms of; overshoot, tracking accuracy and control efforts. Finally, simulation results prove that the proposed control provides an improved dynamic responses and perfect decoupled control in steady and transient states.

Keywords: quadrotor, type-2 Fuzzy Logic Control (T2-FLC), Siding Mode Control (SMC), Memberships functions (MFs))

1. INTRODUCTION

An unmanned aerial vehicle (UAV) is defined as an aircraft which can fly autonomously based on an automation system or can be driven by a pilot at a ground control station. They are mainly appreciated since they can accomplish various missions in the civilian and military areas without putting human lives at risk. Furthermore, the use of UAVs is not restricted only to military and dangerous roles, as it is commonly believed. Moreover, peoples nowadays are showing off using aerial photography [1].

More recently, a growing interest has been shown in this field of UAVs among the research community. Various configurations were proposed. The vehicles with rotary wings have a major advantage compared to those with fixed wings, particularly in an environment, where the capacity of the hovering is useful. The most known UAVs with rotary wings are the conventional helicopters and the quadrotors. This last has certain advantages compared to the conventional helicopters. In fact, it is easier to carry out a hovering with four forces of pushes operating at the same distance from the center of mass as with only one force of push acting on the center of mass. It is highly maneuverable and can be flown safely indoors which makes it well suited for laboratory or hobbyist use. Compared to conventional helicopters, with the large main rotor and tail rotor, the quadrotor does not have the complex swash plate mechanism and are generally low cost and easy to construct [2].

By using four rotors to generate the thrust force, each rotor can be much smaller in diameter. This reduces the amount of kinetic energy during fly, which makes quadrotors much safer [2]. All these factors contribute to making them the rotorcraft of choice for most academic and research purposes [3].

Up to now, many control algorithms for quadrotor unmanned helicopter QUH have been put forward, while everyone has advantages and disadvantages [4]. Algorithms like PD, feedback linearization, Backstepping control and sliding mode control have mainly been applied and proved [5, 6, 16].

It noted that in recent years, the applications of Artifitial intelligent algorithm is extended especially in the Quadrotor field in order to remove or mitigate the drawbacks of conventional controllers. In this context, intelligent algorithms such as Type-2 Fuzzy Logic Control (T2-FLC) is used instead classical controller such as; PI (Proportional Integral) [7, 8].

T2-FLC presents a new class of fuzzy logic systems, and have many advantges over conventional controllers [7], which are do not require an accurate mathematical model and can work with imprecise inputs, can handle non-linearity and are more robust than conventional PI controllers [15].

The main idea which is proposed in this chapter is a comparative study of Quadrotor trajectory tracking using SMC and T2-FLC. The performances criteria based on the study are: system stability, static error in the angle and position, response time and overshoot in transient and steady states in the system.

In this chapter, firstly overview of type-2 fuzzy logic controller Toolbox is presented in section 2, The dynamic model of quadrotor in section 3, Control strategy developed in section 3, Simulation results are demonstrated in section 4. The performance of the proposed controller is discussed in section V.

2. OVERVIEW OF TYPE-2 FUZZY LOGIC CONTROLLER TOOLBOX

2.1. Windwos, Editors, and Viewers of the Type-2-Fuzzy 2.1. Logic Toolbox

Fuzzy inference is a technique that interprets the values in the input vector and, based on user-defined rules, assigns values to the output vector. Using the GUI editors and viewers in the Fuzzy Logic Toolbox, the user can build the rules set, define the membership functions, and analyze the behavior of a fuzzy inference system (FIS). Figure 1-(A) shows the FIS Editor that displays general information about a type-2 fuzzy inference system. It displays actually a menu bar that allows the user to open related GUI tools, open and save systems [12].

2.2. Rule Editor

Figure 1-(B) shows the rule editor that allows the user to view and edit fuzzy rules. On the Rule Editor, there is a menu bar that lets the user to open related GUI tools, and to open and save systems, change the format of the rules and so on. The menus of the Rule Editor are similar to the menus of the other editors.

2.3. Membership Function Editor

The membership function editor shown in Figure 1-(C) allows the user to display and edit the membership functions associated with the input and output variables of the FIS [7].

2.4. Surface Viewer

The Surface Viewer shown in Figure 1-(D) generates a 3-D surface from two input variables and the output of an FIS. The menu bar on the

Surface Viewer allows the user to open related GUI tools, open and save systems, and so on [8-9].

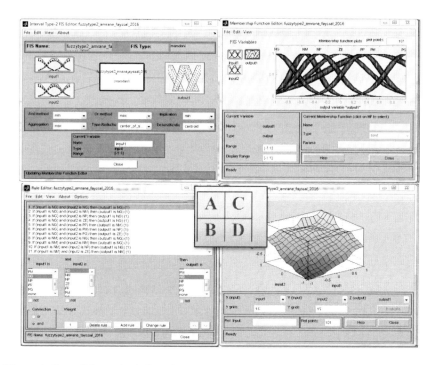

Figure 1. Type-2 fuzzy logic interface; A: Membership function editor of Type-2 fuzzy logic toolbox, B: Rule editor of type-2 fuzzy logic toolbox, C: 2 Inputs-1output Membership function editor of type-2 fuzzy logic Toolbox, D: Surface viewer of type-2 fuzzy logic toolbox.

3. MODEL QUADROTOR DESCRIPTION AND MODELING

A quadrotor UAV is a multirotor aircraft which used four rotors installed on the ends of a symmetrical cross normally out of carbon fiber. The electronics control system is usually placed in the center. The propellers used turn at angular velocities ω_1 ω_2, ω_3 and ω_4 with fixed step. Rotors of the same branch of the cross turn in a direction and the other pair turns in the opposite direction This neutralizes indeed the reactive couples and makes the flight possible without turning out of the command [10-11].

Unlike conventional helicopter which use a tail rotor. The downside in that it requires additional motors that do not contribute to lift. This configuration of pairs moving in opposite directions removes the need for a tail rotor.

The concept of the quadrotor can be seen in Figure 2. The four actuators (Brushless DC motors) produce lift forces F_i proportional to the square of ω_i . and a moment M_i around Z_B axis. M_i is perpendicular to the plan of the appropriate propeller i. Direction of M_i is opposed to the rotational direction of the suitable propeller i. F_T is the total force generated by the propeller system.

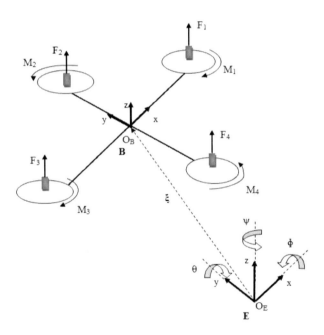

Figure 2. Structure model of quadrotor.

The coordinate frames and free body diagram for the quadrotor are shown in Figure 1. The inertial frame, E, is defined by (O_E, x, y, z). It is usually assumed that the body frame, B(O_B, x, y, z) and the quadrotor center of mass coincide. Rotor 1 is on the positive x axis, 2 on the positive y axis, and the z axis is oriented to complete a right-handed coordinate system.

Type-2 Fuzzy Logic Control for Quadrotor 173

The translational motion is oppossed by the gravity force F_G and the drag force F_D. The torques acting on the quadrotor are: the torque developed by the quadrotor $\tau_{\phi,\theta,\psi}$ the aerodynamic torque friction τ_a and the gyroscopic torque $\tau_g = \tau_{gp} + \tau_{gq}$. ($\tau_{gp}$ and τ_{gq} are respectively propellers gyroscopic torque and the gyroscopic torque due to the quadrotor movement).

We use the Euler angles to model the rotation of the quadrotor in the inertial frame. The community of aeronautics generally uses convention ZYX. The matrix $_B^I R$ allows passing from B to E, it is given by (equation 1):

$$_B^I R = \begin{bmatrix} c\theta c\psi & s\phi s\theta c\psi - c\phi s\psi & c\phi s\theta c\psi + s\phi s\psi \\ c\theta s\psi & s\phi s\theta s\psi + c\phi c\psi & c\phi s\theta s\psi - s\phi c\psi \\ -s\theta & s\phi c\theta & c\phi c\theta \end{bmatrix} \tag{1}$$

Where: c(-), s(-) and t(-) denote in this article cos(-), sin(-) and tan(-), respectively. The position vector $[x\ y\ z]^T$ of the center of mass in the inertial frame is denoted by ξ. $\Omega = [\Omega_x\ \Omega_y\ \Omega_z]^T$ is the vector of the angular speeds of the quadrotor wrt the axes of **B** frame, whose expression in function of Euler's angles speed variation is given by (equation 2) :

$$\Omega = \begin{bmatrix} 1 & 0 & -s\theta \\ 0 & c\phi & c\theta s\phi \\ 0 & -s\phi & c\phi c\theta \end{bmatrix} \begin{bmatrix} \dot{\phi} \\ \dot{\theta} \\ \dot{\psi} \end{bmatrix} \tag{2}$$

Supposing that the quadrotor operate with small angles, the vector Ω can be assimilated to $\begin{bmatrix} \dot{\phi} & \dot{\theta} & \dot{\psi} \end{bmatrix}^T$.

Now, using the Newton's laws about translation and rotational motion, we establish the mathematical model for the quadrotor UAV given by (equation 3).

$$\begin{cases} \ddot{x} = \frac{1}{m}\left[\left(c\phi s\theta c\psi + s\phi s\psi\right)U_z - k_{dx}\dot{x}\right] \\ \ddot{y} = \frac{1}{m}\left[\left(c\phi s\theta s\psi - s\phi c\psi\right)U_z - k_{dy}\dot{y}\right] \\ \ddot{z} = \frac{1}{m}\left[c\phi c\theta\ U_z - k_{dz}\dot{z} - mg\right] \\ \ddot{\phi} = \frac{1}{I_x}\left[\left(I_y - I_z\right)\dot{\theta}\dot{\psi} + U_\phi - k_{d\phi}\dot{\phi}^2 - I_r\bar{\Omega}\dot{\theta}\right] \\ \ddot{\theta} = \frac{1}{I_y}\left[\left(I_z - I_x\right)\dot{\phi}\dot{\psi} + U_\theta - k_{d\theta}\dot{\theta}^2 + I_r\bar{\Omega}\dot{\phi}\right] \\ \ddot{\psi} = \frac{1}{I_z}\left[\left(I_x - I_y\right)\dot{\phi}\dot{\theta} + U_\psi - k_{d\psi}\dot{\psi}^2\right] \end{cases} \tag{3}$$

With: $\bar{\Omega}$ is the signed sum of the angular velocities of the propellers

$$\bar{\Omega} = -\Omega_1 + \Omega_2 - \Omega_3 + \Omega_4 \tag{4}$$

It is important to note that U_z, U_ϕ, U_θ and U_ψ are the control inputs of the system. They are written according to the angular velocities of the four rotors as follows:

$$\begin{cases} U_z = k_F\left(\omega_1^2 + \omega_2^2 + \omega_3^2 + \omega_4^2\right) \\ U_\phi = l.k_F\left(\omega_4^2 - \omega_2^2\right) \\ U_\theta = l.k_F\left(\omega_3^2 - \omega_1^2\right) \\ U_\psi = k_M\left(-\omega_1^2 + \omega_2^2 - \omega_3^2 + \omega_4^2\right) \end{cases} \tag{5}$$

In this section we have established the dynamic model of the quadrotor. It describes the nonlinear behavior, it is an under-actuated system with six outputs {x, y, z, ϕ, θ, ψ} but only four independent inputs $\left\{U_z, U_\phi, U_\theta, U_\psi\right\}$. It is also useful to note that the translation movements depend on the angles.

4. CONTROL STRATEGY

From the notes drawn in the previous section about the quadrotor dynamic model, it is worthwhile to introduce virtual inputs based on the tilt angles (ϕ, θ) as follows:

$$\begin{cases} U_x = (c\phi s\theta c\psi + s\phi s\psi) \\ U_y = (c\phi s\theta s\psi - s\phi c\psi) \end{cases} \quad (6)$$

Now, with six control laws $(U_x, U_y, U_z, U_\phi, U_\theta, U_\psi)$, it is possible to achieve the full control of quadrotor. The control scheme recommended is then, based on two cascade loops. According to the desired values x^d and y^d, the outer control loop calculates desired roll and pitch angles (ϕ^d and θ^d) for the inner one. The benefit of this inner control loop is to calculate corresponding thrusts for desired altitude and attitude, then given as control inputs for the system $(U_z, U_\phi, U_\theta, U_\psi)$. Figure 2 shows the sketch of the proposed control strategy.

In the internal loop and the external loop, the control signals U_ϕ, U_θ, U_ψ, and, U_x, U_y, U_z are getting while applying a T2-FLC.

From (6) it is easy to get desired roll and pitch angles, they are calculated in the external loop (correction block) according to (equation 7):

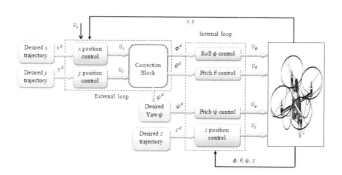

Figure 3. Synoptic scheme of the adopted control system 3.

$$\begin{cases} \phi^d = arcsin\left(U_x \sin\left(\psi^d\right) - U_y \cos\left(\psi^d\right)\right) \\ \theta^d = arcsin\left(\dfrac{U_x \cos\left(\psi^d\right) + U_y \sin\left(\psi^d\right)}{\cos\left(\phi^d\right)}\right) \end{cases} \tag{7}$$

For the state space representation, in this chapter $X = [x_1, \ldots x_{12}]^T$ is considered as the state vector, defined as follows:

$$X = \left[x, \dot{x}, y, \dot{y}, z, \dot{z}, \theta, \dot{\theta}, \phi, \dot{\phi}, \psi, \dot{\psi}\right]^T \tag{8}$$

4.1. Design of Type-2 Fuzzy Logic Controller

The structure of a type-2 Fuzzy Logic System (FLS) is shown in Figure 5. It is actually very similar to the structure of an ordinary type-1 FLS. It is assumed in this chapter that the reader is familiar with type-1 FLSs and thus, in this section, only the similarities and differences between type-2 and type-1 FLSs are underlined.

The fuzzifier shown in Figure 5, as in a type-1 FLS, maps the crisp input into a fuzzy set. This fuzzy set can be a type-1, type-2 or a singleton fuzzy set. In singleton fuzzification, the input set has only a single point of nonzero membership. The singleton fuzzifier is the most widely used fuzzifier due to its simplicity and lower computational requirements.

The shaded region in Figure 4a is the FOU for a type-2 fuzzy set. The primary memberships, J_{x1} and J_{x2}, and their associated secondary membership functions $\mu_{\tilde{A}}(x_1)$ and $\mu_{\tilde{A}}(x_2)$ are shown at the points x_1 and x_2. The upper and lower membership functions, $\overline{\mu}_{\tilde{A}}(x)$ and $\underline{\mu}_{\tilde{A}}(x)$, are also shown in Figure 4.a. The secondary membership functions, which are interval sets, are shown in Figure 4.b. [12].

The third dimension of type-2 fuzzy sets decides secondary membership function and the FOU decides the range of uncertainty, together provides additional degree of freedom in the design to compensate various uncertainties [13].

The type2 fuzzy controller utilized in this work has two inputs and one output. The membership functions are defined in Figure 6 (A & B).

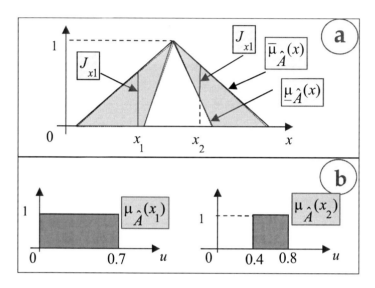

Figure 4. (a) The FOU for a type-2 fuzzy set, (b) The secondary membership functions.

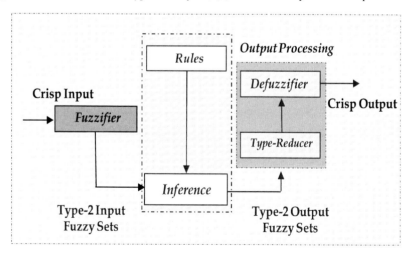

Figure 5. The structure of a type-2 Fuzzy Logic System.

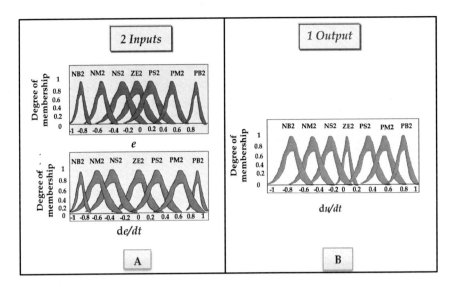

Figure 6. T2-FLC Membership functions (A: 02 inputs and B: 01 output).

The type 2 fuzzy rule base consists of a collection of linguistic rules of the form [14]:

Rule 1: if $S_{1,2}$ is NB2, and $S_{1,2}$ is NB2 then $dU_{1,2}$ is NB2.
Rule 2: if $S_{1,2}$ is NM2, and $S_{1,2}$ is NB2 then $dU_{1,2}$ isNB2.
Rule 3: if $S_{1,2}$ is NS2, and $S_{1,2}$ is NG2 then $dU_{1,2}$ is NS2.
Rule 4: if $S_{1,2}$ is PB2, and $S_{1,2}$ is PB2 then $dU_{1,2}$ isPB2.

4.2. Design of Sliding Mode Control

A nonlinear control "Sliding Mode Control" has been applied on a model of the X4-flyer with the parameters given in Table 1. Where s is the SMC, λ and $\xi > 0$ coefficients of the SMC.

Type-2 Fuzzy Logic Control for Quadrotor

$$
\left\{
\begin{aligned}
U_x &= \frac{m}{U_z}\left[\ddot{x}_1^d - \lambda_x \dot{e}_x - \xi_x s_1 - \mu_x sign(s_1) + \frac{k_{dx}}{m}x_2\right] \\[4pt]
U_y &= \frac{m}{U_z}\left[\ddot{x}_3^d - \lambda_y \dot{e}_y - \xi_y s_2 - \mu_y sign(s_2) + \frac{k_{dy}}{m}x_4\right] \\[4pt]
U_z &= \frac{m}{\cos\varphi\cos\theta}\left[\ddot{x}_5^d - \lambda_z \dot{e}_z - \xi_z s_3 - \mu_z sign(s_3) + \frac{k_{dz}}{m}x_6 + g\right] \\[4pt]
U_\varphi &= I_x\left[\ddot{x}_7^d - \lambda_\varphi \dot{e}_\varphi - \xi_\varphi s_4 - \mu_\varphi sign(s_4) - \frac{\left(I_y - I_z\right)}{I_x}x_{10}x_{12} + \frac{k_{d\varphi}}{I_x}x_8^2 + \frac{I_r}{I_x}\bar{\Omega}x_{10}\right] \\[4pt]
U_\theta &= I_y\left[\ddot{x}_9^d - \lambda_\theta \dot{e}_\theta - \xi_\theta s_5 - \mu_\theta sign(s_5) - \frac{\left(I_z - I_x\right)}{I_y}x_8 x_{12} + \frac{k_{d\theta}}{I_y}x_{10}^2 - \frac{I_r}{I_y}\bar{\Omega}x_8\right] \\[4pt]
U_\psi &= I_z\left[\ddot{x}_{11}^d - \lambda_\psi \dot{e}_\psi - \xi_\psi s_6 - \mu_\psi sign(s_6) - \frac{\left(I_x - I_y\right)}{I_z}x_8 x_{10} + \frac{k_{d\psi}}{I_z}x_{12}^2\right]
\end{aligned}
\right. \tag{9}
$$

5. SIMULATION RESULTS

In order to verify the effectiveness of the proposed controller, flying tests were carried out in Matlab/Simulink® 2017a, the physical parameters of quadrotor aerial robot are described in details in Table. 2.

Table 1. Reference Trajectory

Variable	Value	Time (s)
	[2, 2, 2]	[0 10]
	[1, 2, 2]	[10 20]
	[2, 1, 2]	[20 30]
[xd, yd, zd]	[2, 1, 2]	[30 40]
	[2, 2, 2]	[40 50]
	[0.5Sin(0.5t) + 2, 0.5Cos(0.5t) + 2, -0.01t]	[50 100]
Ψ	[Pi/6 0]	[0 50 100]

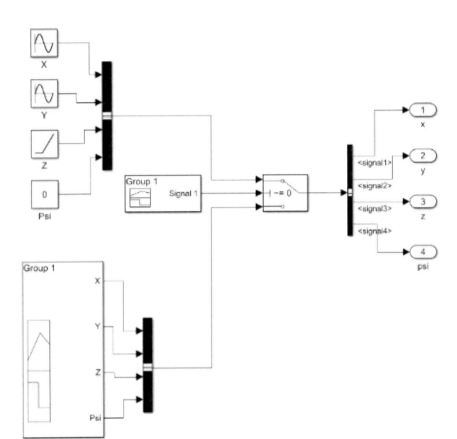

Figure 7. Trajectory block.

The desired trajectory chosen is to make a vertical track off of 2 (m) altitude, then, a square trajectory in the air. After that, a 2 (m) surface square from the point (2,2,2) to the last point of the square. Finally, the quadrotor must make a Helical trajectory from the point (2,2,2). With 0 (rad) ψ angle.

Figure 10 shows Quadrotor's path in Three-dimensional space (3D) of T2-FLC and SMC. It shows a good performance towards stability and path tracking even after change of trajectory, which explains the efficiency of the proposed control strategy developed in this chapter.

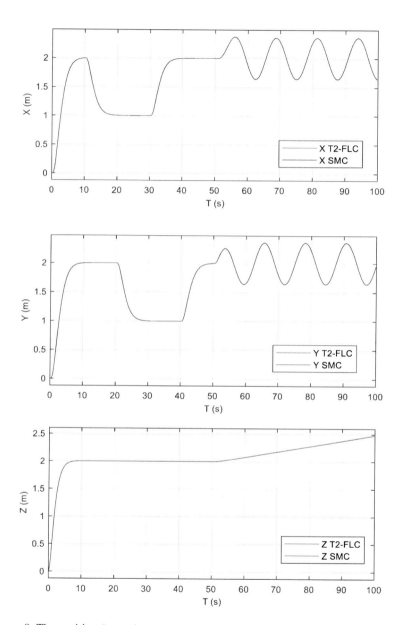

Figure 8. The position (x, y, z).

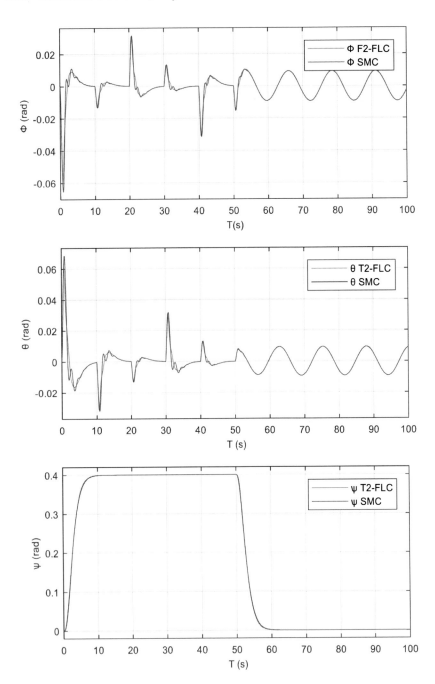

Figure 9. The angles (ϕ, θ, ψ).

Type-2 Fuzzy Logic Control for Quadrotor 183

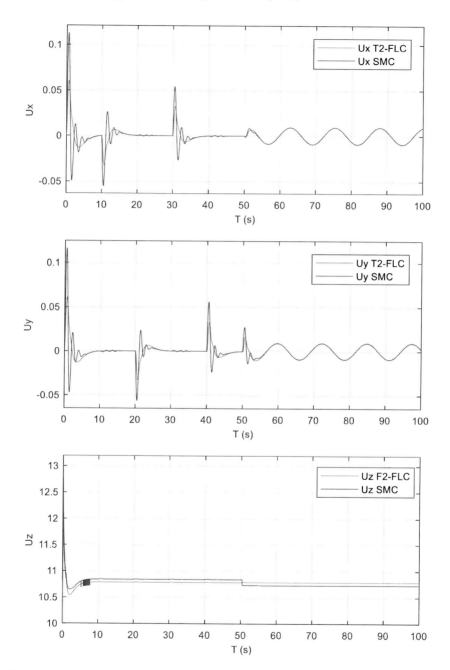

Figure 10. The controllers (U_x, U_y, U_z).

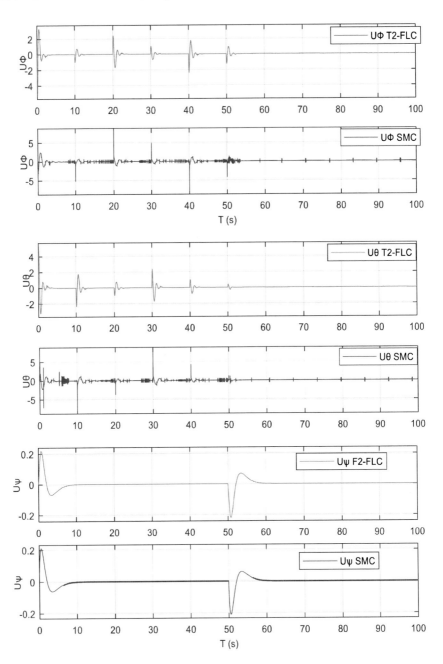

Figure 11. The controllers (U_ϕ, U_θ, U_ψ).

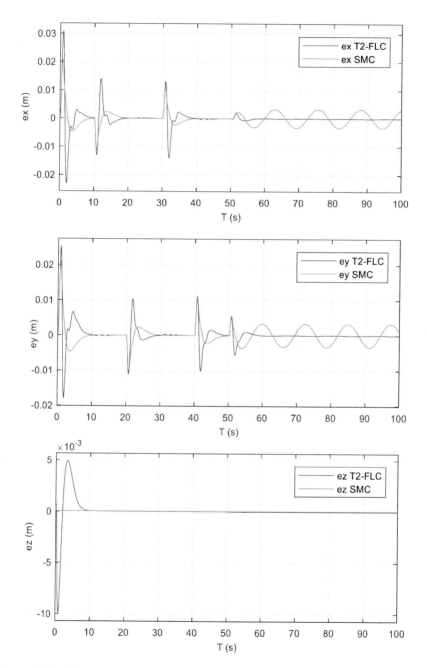

Figure 12. Tracking error in position.

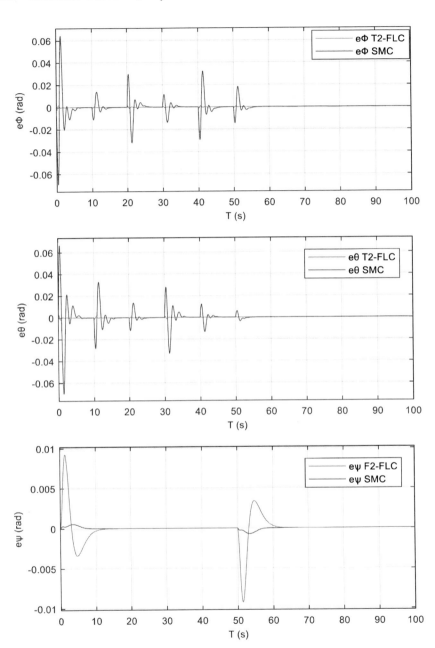

Figure 13. Tracking errors in attitude.

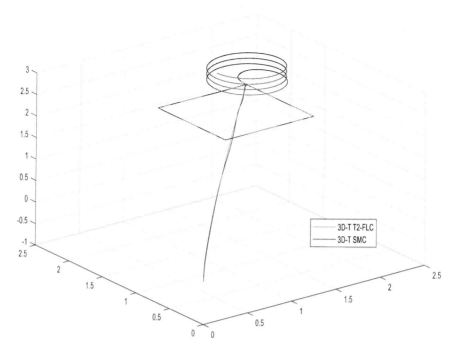

Figure 14. Quadrotor's path in set-point angle and position

CONCLUSION

In this chapter a comparative analysis study between a Sliding Mode Control (SMC) and Type-2 Fuzzy Logic Control (T2-FLC) is proposed for 6 DOF Quadrotor to solve the trajectory tracking and stability problems. According to the obtained results the study carried out through simulations shown that both techniques manage to control the action of the quadrotor. In the SMC controller many adjustments were necessary to the gains that resulted in a stability of the system, whereas with the T2-FLC a high performance and stability are noted under deffirent trajectory profiles againts SMC. It noted that a remarkable undulations and harmonics are shown in steady state espacially using SMC, and they influence in the quadrotor stability system.

Future work intends to implement the simulation in real time wiht a hybrid T2-FLC and Artifitial Neural Networks (ANN), as well as verify

188 *Gherouat Oussama, Fayssal Amrane and Abdelouahab Hassam*

the feasibility of other Artificial Intelligence techniques. Also, for better precision, the optimization algorithms are necessary in order to calculate precisely the gains values of controllers.

APPENDIX

Table 2. Physical Parameters of Quadrotor Aerial Robot

Symbol	Value	Physical Significance
I_x	1.22	Quadrotor moment of inertia around X axis (N.S²/rad)
I_y	1.22	Quadrotor moment of inertia around Y axis (N.S²/rad)
I_z	2.2	Quadrotor moment of inertia around Z axis (N.S²/rad)
I_r	0.2	Total rotational moment of inertia around the rotor axis (N.S²/rad)
k_F	5	Lift factor (N/rad/s)
k_M	2	Drag factor (N.m/rad/s)
l	0.21	Arm length (m)
m	1.1	Total mass of the quadrotor (Kg)
g	9.81	Acceleration due to gravity (m/s²)
k_{dx}	0.1	Translational drage coefficient according to X axies (Ns/m)
k_{dy}	0.1	Translational drage coefficient according to Y axies (Ns/m)
k_{dz}	0.1	Translational drage coefficient according to Z axies (Ns/m)
$k_{d\phi}$	0.12	Rotational drage coefficient, Roll movement (N/rad/s)
$k_{d\theta}$	0.12	Rotational drage coefficient, Pitch movement (N/rad/s)
$k_{d\psi}$	0.12	Rotational drage coefficient, Yaw movement (N/rad/s)

REFERENCES

[1] Castillo, P., R. Lozano, A.E. Dzul, *"Modeling and Control of Mini-Flying Machines,"* France, 2004.

[2] Leishman, J.G., *"Principles of Helicopter Aerodynamics."* New York, NY: Cambridge University Press (2000).

[3] Siciliano, B., O. Khatib, F. Groen, *"Robotics, Vision and Control,"* Springer, 2011, Vol. 73.

[4] Li, Y., S. Song, "A Survey of Control Algorithms for Quadrotor Unmanned Helicopter," *IEEE fifth International Conference on Advanced Computational Intelligence* (ICACI), China, 2012.

[5] Matouk, D., O. Gherouat, F. Abdessemed and A. Hassam, "Quadrotor position and attitude control via backstepping approach," *2016 8th International Conference on Modelling, Identification and Control* (ICMIC), Algiers, 2016, pp. 73-79.

[6] Gherouat, Oussama & Matouk, Djihad & Hassam, Abdelouahab & Abdessemed, Foudil. (2017). Sliding Mode Control for a Quadrotor Unmanned Aerial Vehicle. *J. Automation & Systems Engineering.* 10. 150-157.

[7] Amrane, Fayssal and Azeddine Chaiba. *Fuzzy Control Systems: Design, Analysis and Performance Evaluation,* Chapter 1. Type2 Fuzzy Logic Control: Design and Application in Wind Energy Conversion System based on DFIG via Active and Reactive Power Control, pp. 1-35.

[8] Amrane, Fayssal, Azeddine Chaiba and Bruno Francois Application of Adaptive T2FLC in Stator Active and Reactive Power Control WECS based on DFIG via Sub/Super- Synchronous Modes, *Conférence nationale des Jeunes Chercheurs en Génie Electrique* (JCGE' 2017), Arras, France.

[9] Amrane, Fayssal, Azeddine Chaiba and Bruno Francois, "Suitable power control based on type-2 fuzzy logic for wind-turbine dfig under hypo-synchronous mode fed by multi-level converter," 2017 *5th International Conference on Electrical Engineering - Boumerdes (ICEE-B)*, Boumerdes, 2017, pp. 1-6.

[10] Lantos, B., L. Màrton, "Nonlinear Control of Vehicles and Robots," *Advances in Industrial Control*, Spronger, 2011.

[11] Castillo, P., R. Lozano, A.E. Dzul, "Modelling and Control of Mini-flying Machines," *A Advances in Industrial Control*, Spronger, 2005.

[12] Muzeyyen Bulut Ozek; Zuhtu Hakan Akpolat. *A software tool: type-2 fuzzy logic toolbox JWPI.* 2008, vol, 22.

[13] Raju, Krishnama S., G. N. Pillai. "Design and Implementation of Type-2 Fuzzy Logic Controller for DFIG-Based Wind Energy Systems in Distribution Networks," *IEEE TSE*. 2016, vol, 7.
[14] Suganthia, L., S. Iniyan and Anand A. Samuel. *Applications of fuzzy logic in renewable energy systems – A review JRSER*. 2015, vol, 48.
[15] Harvey, Terrell, Dallas Mullins, "*Fuzzy Modeling and Control: Methods, Applications and Research,*" Nova Science Publishers, 2018 - 163 pages.
[16] Matouk, D., O. Gherouat, F. Abdessemed and A. Hassam, "Robust Control of a Quadrotor Unmanned Aerial Vehicle," *International Conference on Technological Advances in Electrical Engineering* (ICTAEE' 16), October 2016, Algiers.

BIOGRAPHICAL SKETCH

Oussama Gherouat

Oussama Gherouat was born in Setif Algeria, in 1988. He received his License and Master degrees in electrical engineering from University of Setif-1, Algeria, and University of Batna, Algeria in 2011 and 2013, respectively. Actually, he is member of LSI (Laboratory of Intelligent Systems). He is currently working toward his Ph.D. in the Department of Electronic at the University of Setif-1 His specific research interests are

Type-2 Fuzzy Logic Control for Quadrotor 191

Computational Intelligence Methods, Nonlinear Control Theory, Sliding Mode Control, Backstepping control, Fuzzy logic, Mechatronics and unmanned aerial Vehicles and Shunt Active Power Filter.

Education:

- Ph.D. student in the department of Electronic at university of Ferhat Abbas Setif 1 ALGERIA LSI (Laboratory of Intelligent Systems)
- Master degree in control of industrial systems, Department of Electrical Engineering, University of Batna, Algeria, 2013.
- License degree in Automatic Control, Department of Electrical Engineering, University of Setif-1, Algeria, 2011.

Publications:

1. Djihad Matouk, Oussama Gherouat, Foudil Abdessemed and Abdelouahab Hassam. (2016, November). Quadrotor position and attitude control via backstepping approach. In *Modelling, Identification and Control (ICMIC), 2016 8th International Conference on* (pp. 73-79). IEEE

2. Oussama Gherouat, Djihad Matouk, Abdelouahab Hassam, and Foudil Abdessemed (*November 14-15, 2016)*, Sliding Mode Control of Quadrotor Unmanned Aerial Vehicle, 1st National Conference on Electronics and Electrical Engineering (NCEEE'16)

3. Oussama Gherouat, Djihad Matouk, Abdelouahab Hassam, and Foudil Abdessemed. (2016, October),Combined Backstepping and Enhanced PD Control Design for Position and Attitude Stabilization of an UAV Quadrotor, The International Conference on Electrical Sciences and Technologies in Maghreb October 26th- 28th, 2016, Marrakesh, Morocco

4. Oussama Gherouat, Djihad Matouk, Abdelouahab Hassam, and Foudil Abdessemed, (October 2016) Robust Control of a Quadrotor Unmanned Aerial Vehicle, *International Conference on Technological Advances in Electrical Engineering (ICTAEE'16.),*

5. Oussama Gherouat, Djihad Matouk, Abdelouahab Hassam, and Foudil Abdessemed, Sliding Mode Control for a Quadrotor

Unmanned Aerial Vehicle, J. Automation & Systems Engineering 10-3 (2016): 150-157

Professional Experience:
- Assistant Professor, (Practical work in data processing) Faculty of Economic Sciences, Business and Management, Department of Management, Setif1 University,2018.
- Assistant Professor (Practical work in applied electronics) Department of Electrical Engineering, University of Setif1, Algeria, 2017.
- Assistant Professor (Math) polymer engineering department, University of Setif1, Algeria, 2016.

Current Research Interests:
Computational Intelligence Methods, Nonlinear Control Theory, Sliding Mode Control, Backstepping control, Fuzzy logic, Mechatronics and unmanned aerial Vehicles and Shunt Active Power Filter

Abdelouahab Hassam

Abdelouahab Hassam was born in Setif, Algeria. He received the B.Sc.degree in Electronics Engineering, in 1983, from the National Polytechnic School of Algiers (ENPA), Algeria, in 2007, and he received his magister and his Doctoral degree from University of Setif, Algeria, where he is a Professor and the head Department of Electronics Engineering. His specific research interests are robotics, information theory, signal processing, and sensors.

Affiliation:
Department of Electronics Engineering, University of Setif 1, 19000 Setif, Algeria.

Education:
- PhD in trajectory planning and control of movements of a robot Department of Electronics Engineering, University of Setif-1, Algeria, 2007.
- Magister degree in microcomputer. Department of Electronics Engineering, University of Setif-1, Algeria, 1993.
- B.Sc. degree in microcomputer, Department of Electronics Engineering, National Polytechnic School of Algiers (ENPA), Algeria.1983.

Address:
Department of Electronics Engineering, Setif-1 University, 19000, Setif-1, Algeria.

Research and Professional Experience:

From 1983
Professor in Electricity; Mechanics, Atomic and Nuclear Physics; Electronic measurements; Physics of Semiconductors General Electronics; Servo Systems; Power Electronics; Reliability and Instrumentation; Industrial Robotics. Faculty of Science and Technology, at Setif-1 University.

His Current Research Interests Include:
- Robotics, information theory, signal processing, and sensors., nonlinear control and Artificial intelligence in robot control.

194 *Gherouat Oussama, Fayssal Amrane and Abdelouahab Hassam*

- **Journal Papers**
  1. Boubezoula Mabrouk, Abdelouahab Hassam and Oussama Boutalbi "Robust-flatness Controller Design for Differencially Driven Wheeled Mobile Robot, International Journal of Control, Automation and Systems, ICROS, KIEE and Springer, 2018
  2. Fawzi Srairi, Lamir Saidi and Abdelouahab Hassam "Modeling Control and Optimization of a New Swimming Microrobot Using Flatness-Fuzzy-Based Approach for Medical Applications" Arabian Journal for Science and Engineering, pp. 3249–3258 Springer, 2018
  3. Bedaouche Fatah, Abdelouahab Hassam and Boubezoula Mabrouk "Stability Guaranteed and Simulation of héliostat on Testing-Ground" e journal.ktu.lt, vol. 23, no. 5, 2017
  4. Miloud Hamani and Abdelouahab Hassam "Improving mobile robot navigation by combining fuzzy reasoning and virtual obstacle algorithm", Journal-of-Intelligent-and-FuzzySystems, vol. 30, no. 3, pp. 1499-1509, 2016

- **International Conferences in: 2015, 2016 and 2017**
  1. Djihad Matouk, Oussama Gherouat, Foudil Abdessemed and Abdelouahab Hassam.(2016, November). Quadrotor position and attitude control via backstepping approach. In *Modelling, Identification and Control (ICMIC), 2016 8th International Conference on* (pp. 73-79). IEEE
  2. Oussama Gherouat, Djihad Matouk, Abdelouahab Hassam, and Foudil Abdessemed (*November 14-15, 2016)*, Sliding Mode Control of Quadrotor Unmanned Aerial Vehicle, 1st National Conference on Electronics and Electrical Engineering (NCEEE'16)
  3. Oussama Gherouat, Djihad Matouk, Abdelouahab Hassam, and Foudil Abdessemed. (2016, October),Combined Backstepping and Enhanced PD Control Design for Position and Attitude Stabilization of an UAV Quadrotor, The International Conference on Electrical Sciences and

Technologies in Maghreb October 26th- 28th, 2016, Marrakesh, Morocco

4. Oussama Gherouat, Djihad Matouk, Abdelouahab Hassam, and Foudil Abdessemed, (October 2016) Robust Control of a Quadrotor Unmanned Aerial Vehicle, *International Conference on Technological Advances in Electrical Engineering (ICTAEE'16.),*

5. Oussama Gherouat, Djihad Matouk, Abdelouahab Hassam, and Foudil Abdessemed, Sliding Mode Control for a Quadrotor Unmanned Aerial Vehicle, J. Automation & Systems Engineering 10-3 (2016): 150-157

In: Advances in Engineering Research
Editor: Victoria M. Petrova

ISBN: 978-1-53616-092-5
© 2019 Nova Science Publishers, Inc.

Chapter 6

USING 2D SPATIAL FILTERING TECHNIQUES TO SIMULATE SURFACE SINGLE FIBRE ACTION POTENTIAL

Noureddine Messaoudi[1,2,*], Raïs El'hadi Bekka[2]
and Samia Belkacem[1]

[1]Department of Electrical Systems Engineering, Faculty of Engineering
Sciences, Université de Boumerdes
Boumerdes, Algeria
[2]Lab. LIS, Electronics Department, Faculty of Technology,
Université de Sétif 1, Sétif, Algeria

ABSTRACT

Surface single fibre action potential (SFAP) is the main component of the surface electromyographic (EMG) signal. In this chapter, an analytical approach called "two-dimensional (2D) spatial filtering techniques" was used to simulate the surface SFAP signal. The volume conductor and the detection system were considered as 2D spatial filter, its input signal was the current density source and its output signal was the potential detected on the skin surface. The volume conductor was

---

* Corresponding Author's Email: n.messaoudi@univ-boumerdes.dz.

198 *Noureddine Messaoudi, Raïs El'hadi Bekka and Samia Belkacem*

considered as a planar, space-invariant, multilayer (non-homogeneous) and anisotropic medium constituted by muscle, fat and skin tissues. The detection system was obtained by combining the spatial filter, size and shape of the electrodes. One (1D) and two (2D) spatial filters were investigated. Rectangular, circular, elliptical and annular shaped electrodes were used to detect the SFAP signal. Two current density (impulsive and analytical) sources were used for the generation of surface SFAP signal and a comparison study was made. We showed that the surface SFAP generated by an impulsive source is the outcome of the spatial phenomena and the one generated by an analytical source is due to the diameter and to the intracellular conductivity of the fibre.

Keywords: electromyography, simulation, spatial filtering, volume conductor

INTRODUCTION

Modelling of the surface electromyography (sEMG) is very useful for studying the relationships between sEMG signal features and the underlying physiological processes [1], in the interpretation of experimental results [2], for the test and validation of new algorithms in sEMG signal [3], in the choice of detection system [4] and for didactic purposes [5]. The sEMG signal is the sum of the trains of motor-unit potentials and the motor-unit potential is the sum of the action potentials of the muscle fibres belonging to all the active motor units. Therefore, simulation of the sEMG signal can be based on a model of the single muscle fibre action potential (SFAP).

Many surface SFAP models signal have already been proposed [6-9]. The difference between these models depends mainly in description of the volume conductor. The volume conductor was described as a space invariant (the surface potentials detected at different electrode locations along fibre direction do not change shape) [6-8] or not space invariant in the direction of propagation [10], [11-13] system. Analytical [6], [14] and numerical [15-17] approaches were used for the description of the electrical properties separating the muscle fibres and the detection system. The volume conductor was studied in Cartesian coordinates [7], and in

Using 2D Spatial Filtering Techniques to Simulate Surface ... 199

cylindrical coordinates [9], [18]. The main steps of SFAP modelling are [15], [19]: 1) the description of the source, i.e., the modelling of the generation of two intracellular action potentials (IAPs) at the neuromuscular junction, their propagations in two opposite directions along the muscle fibre and their extinctions at the right and left tendons, 2) the mathematical description of the volume conductor, and 3) the modelling of the detection system, i.e., the modelling of the spatial filters used for surface SPAF signal detection, shape and size of the electrodes, the inter-electrode distance. The volume conductor effect is described as a two-dimensional spatial filtering. Electrodes of the same shape are simulated, forming structures which are described as spatial filters.

In this book chapter, the SFAP was simulated using the model proposed by Farina et al., [8]. The volume conductor was described as a multilayer and anisotropic medium constituted by muscle, fat and skin tissues.

We showed that the shape of the simulated surface SFAP signal significantly depends on anatomical, physiological and detection system parameters.

DESCRIPTION METHODS

In the 2D spatial filtering techniques, the volume conductor and the detection system are considered as a 2D spatial filter where its input signal is the Fourier transform of the current density source and its output signal is the surface SFAP signal. In addition, the volume conductor (anatomical and physiological parameters) and the detection system (spatial filter, electrodes shape and size and inter-electrode distances) were described analytically by transfer functions in the 2D spatial frequency domain. In this method the volume conductor (anatomical and physiological parameters) and the detection system (spatial filter, electrodes shape and size and inter-electrode distances) were described mathematically by transfer functions in the 2D spatial frequency domain.

SOURCE DESCRIPTION

Intracellular action potential (IAP) is generated at the neuromuscular junction (NMJ), propagates along the muscle fibres in opposite directions and extinguishes at the right and the left tendons, respectively (Figure 2). The IAP is mathematically described in space domain by the transmembrane potential proposed by Rosenfalck [8]:

$$
V_m(z) = \begin{cases} Az^3 e^{-z} + B, & z > \\ 0, & z \leq 0 \end{cases} \tag{1}
$$

where $A = 96\ mV.mm^{-3}$ and $B = -90\ mV$.

The transmembrane current density source is proportional to the second derivative of the transmembrane potential [16]:

$$
I_m(z) = \frac{\sigma_i \pi (d_1)^2}{4} \frac{d^2 V_m(z)}{dz^2} \tag{2}
$$

where σ_i is the intracellular conductivity of the muscle fibre, d_1 is the muscle fibre diameter and $V_m(z)$ is the intracellular action potential.

VOLUME CONDUCTOR DESCRIPTION

The volume conductor was modelled as a multilayered medium constituted by muscle tissue (anisotropic), and fat and skin tissues (isotropic) as shown in Figure 1. The transfer function of the conductor volume described in a Cartesian coordinate system is expressed in the 2D spatial frequency domain by [8]:

Using 2D Spatial Filtering Techniques to Simulate Surface ... 201

$$H_{vc}(K_x, K_z, y_0) = \frac{2e^{-K_{ya}|y_0|}}{\sigma_{mt}(1+R_c)\cosh[K_y(h_1+d)]\alpha(K_y(h_1+d))+(1-R_c)\cosh[K_y(h_1-d)]\alpha(K_y(h_1-d))}$$

(3)

where

d: thickness of the skin layer,

h_1 :thickness of the fat layer,

y_0 : depth of the fibre,

$$R_c = \frac{\sigma_s}{\sigma_f}, \quad R_m = \frac{\sigma_f}{\sigma_{mz}}, \quad R_a = \frac{\sigma_{ml}}{\sigma_{mz}}, \quad \alpha(K_y) = K_{ya} + R_m K_y tgh(K_y),$$

$$K_y = \sqrt{K_x^2 + K_z^2}, \quad K_{ya} = \sqrt{K_x^2 + R_a K_z^2}$$

with K_x and K_z are the spatial angular frequencies in the transversal and longitudinal directions with respect to the muscle fibres.

σ_{ml} and σ_{mt} are the conductivities of the muscle layer in the longitudinal and transversal directions with respect to the muscle fibres direction (see table 1).

DETECTION SYSTEM DESCRIPTION

According to the spatial filtering techniques, the detection system is the combination of the spatial filter and electrodes shape and size [8]. It was composed by a 1D and 2D spatial filters with physical electrodes (electrodes with dimensions) and of rectangular, circular, elliptical and annular shapes [8].

The studied detection systems were:

- The monopolar, Longitudinal Single (LSD)and Double Differential (LDD), Transversal Single (TSD) and Double (TDD) Differential,for one dimensional spatial filter,
- Normal Double Differential (NDD), Inverse Binomial of order two (IB2) and the Inverse Rectangle (IR) for the two dimensional spatial filters. The 2D spatial filters (NDD, IR and IB2) are generally used in image processing [20].
- The transfer function in the 2D spatial frequency domain of systems is expressed in case of point electrodes by [8]:

$$H_{sf}(K_x, K_z) = \sum_{i=-m}^{n-1} \sum_{r=-\omega}^{h-1} a_{ir} e^{-jK_x id_x} e^{-jK_z rd_z} \tag{4}$$

where

m and n are respectively the number of columns right and left to centre of the electrodes array; ω and h are respectively the number of rows above and below to the centre of the electrodes array; a_{ir} are the weights given to the electrodes; d_x and d_z are the inter-electrodes distances in the x and z directions, respectively.

The electrodes shape and size are also described by transfer functions in the 2D spatial frequency domain. Four electrode shapes were considered; circular, rectangular elliptical electrodes as defined in [8] and concentric-ring electrodes as defined in [21]:

$$H_{cir}(K_x, K_z) = \begin{cases} 2\dfrac{J_1\left(r\sqrt{(K_x)^2 + (K_z)^2}\right)}{r\sqrt{(K_x)^2 + (K_z)^2}} & \text{if } (K_x, K_z) \neq (0,0) \\ 1 & \text{if } (K_x, K_z) = (0,0) \end{cases} \tag{5}$$

r is the radius of the circular electrode.

$$H_{rect}(K_x, K_z) = \sin c\left(\frac{K_x a}{2\pi}\right) \sin c\left(\frac{K_z b}{2\pi}\right) \tag{6}$$

a and *b* are the axes of the rectangular electrode.

$$H_{ellip}(K_x,K_z) = \begin{cases} 2\dfrac{j_1\left(\sqrt{(aK_x)^2+(bK_z)^2}\right)}{\sqrt{(aK_x)^2+(bK_z)^2}} & \text{if } (K_x,K_z) \neq (0,0) \\[4mm] 1 & \text{if } (K_x,K_z) = (0,0) \end{cases} \tag{7}$$

a and *b* are the short and the long axes of the elliptical electrode.

$$H_{ring}(K_x,K_z) = \begin{cases} \dfrac{2}{(r_2^2 - r_1^2)}\left[r_2\dfrac{J_1(r_2 K_y)}{K_y} - r_1\dfrac{J_1(r_1 K_y)}{K_y} \right] & \text{if } K_y \neq 0 \\[4mm] 1 & \text{if } K_y = 0 \end{cases} \tag{8}$$

r_1 and r_2 are respectively internal and external radiuses of the ring; $K_y = \sqrt{K_x^2 + K_z^2}$; $J_1(h)$ is the first order Bessel function of the first kind [8].

For physical electrodes, the detection system transfer function is the product of the spatial filter transfer function and the transfer function described the shape and size of the detection electrodes:

$$H_{sys\_det}(K_x,K_z) = \sum_{i=-m}^{n-1} \sum_{r=-\omega}^{h-1} a_{ir} H_{size}^{ir}(K_x,K_z) e^{-jK_x i d_x} e^{-jK_z r d_z} \tag{9}$$

where $H_{size}^{ir}(K_x,K_z)$ is the transfer function related to the shape and size of the electrode (*ir*).

Computation of the Surface SFAP Signal

For a space invariant volume conductor (as in our case), the potential detected on the skin surface in the 2D spatial frequency domain is obtained by multiplying the product of the global transfer function $H_{glo}(K_x,K_z)$ and the Fourier transform of the current density source $I(K_z)$.

$$\phi(K_x, K_z) = I(K_z) H_{glo}(K_x, K_z) \qquad (10)$$

$H_{glo}(K_x, K_z)$ is obtained by multiplying the volume conductor transfer function $H_{vc}(K_x, K_z)$ with the detection system transfer function $H_{syst\_det}(K_x, K_z)$:

$$H_{glo}(K_x, K_z) = H_{vc}(K_x, K_z) H_{syst\_det}(K_x, K_z) \qquad (11)$$

In the 2D spatial domain, the potential detected on the skin surface is the convolution product of the impulse response corresponds to the global transfer function $H_{glo}(K_x, K_z)$ and the current density source $I_m(z)$.

$$\varphi(x, z) = I_m(z) * IFFT2(H_{glo}(K_x, K_z)) \qquad (12)$$

IFFT2 indicate the two dimensional inverse Fourier transform.

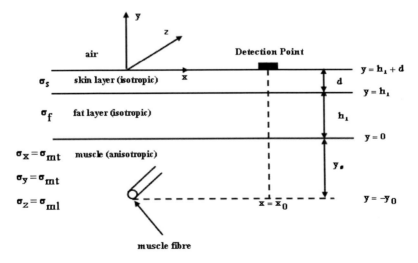

Figure 1. Three layer planar volume conductor constituted by muscle (anisotropic), fat (isotropic) and skin (isotropic) layers. x and y directions are transversal with respect to the z direction of the fibres. The volume conductor is infinite in x and z directions and semi-infinite in y direction [8].

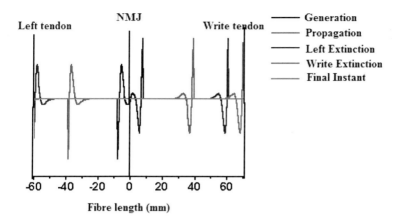

Figure 2. Generation, propagation and extinction of the current density source. The source is generated at the neuromuscular junction (JNM), propagates toward the muscle fibre and extinguishes at the right and the left tendons, respectively.

MODELLING PARAMETERS DESCRIPTION

Table 1. Anatomical, physiological and detection system parameters used to simulate the surface SFAP signal [5]

Parameter	Description	Value
y_0	Depth of the fibres in the muscle	2 mm
h	Thickness of the fat layer	3 mm
d	Thickness of the skin layer	1 mm
σ_s	Conductivity of the skin layer	0.4 S/m
σ_f	Conductivity of the fat layer	0.02 S/m
σ_{ml}	Muscle conductivity in the longitudinal direction	0.5 S/m
σ_{mt}	muscle conductivity in the transversal direction	0.1 S/m
w	Weights given to electrodes	With respect to the spatial filter
$d_x = d_z$	Inter-electrode distances	20 mm
θ	Angle of inclination	-

The simulated surface SFAP was implemented in MATLAB according to method description below. The duration of the simulated signal was $40 ms$ at a sampling frequency of $512 Hz$. The model's basic parameters are shown in Table 1.

RESULTS

Figure 3 shows the transfer functions of two planar volume conductors, one composed of a single layer (muscle) and the other composed of three layers (muscle, fat and skin). This result states that increasing the number of layers of the volume conductor reduces the width of the transfer function.

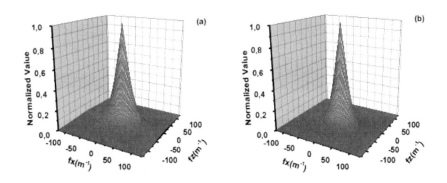

Figure 3. Absolute normalized value of the multilayer planar volume conductor transfer function when it is constituted by: (a) muscle layer and (b) muscle, fat and skin layers.

The configuration of the electrodes enables to distinguish two types of spatial filters, one dimensional and two dimensional detection systems. It was shown in [8] that one dimensional spatial filters have anisotropic transfer functions (non invariant to rotation) and 2D spatial filters have isotropic transfer functions (rotation invariant). These results are confirmed by those obtained in this work as shown in Figure 4.

Using 2D Spatial Filtering Techniques to Simulate Surface ...

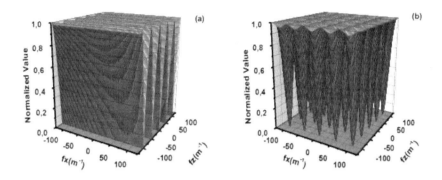

Figure 4. Absolute normalized value of the spatial filter transfer functions in the case of: (a) transversal double differential (TDD) and (b) inverse binomial of order two (IB2) configurations. The inter-electrode distances are $dx = dz = 20$ mm.

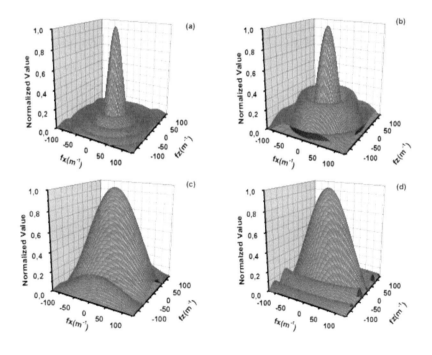

Figure 5. Absolute normalized values of the transfer functions describing the shape and dimensions of the: (a) circular (radius of 15 mm), (b) concentric ring (internal and external radius of 5 mm and 10 mm, respectively), (c) rectangular (axes of 7 mm and 15 mm, respectively) and (d) elliptical (short and large axes of 4.5 mm and 12 mm, respectively) electrodes.

Figure 5(a) and (b) show that the transfer functions of the circular and concentric-ring electrodes are isotropic. Figure 5(c) and (d) show that the transfer functions of the rectangular and elliptical electrodes are anisotropic. The rotation of the rectangular and elliptical electrodes on the skin was simulated by the rotation of their transfer functions and the results are shown in Figure 6 which further shows the anisotropy of the two types of electrodes. So, the isotropy or the anisotropy of a detection system is related to the isotropy or anisotropy of spatial filter and electrode shape at the same time. Figure 7(a) and 7(b) show the transfer function of the detection system constituted by the TDD spatial filter which is an anisotropic filter and the transfer function of the detection system constituted by a grid of circular electrodes which is also an anisotropic. Indeed, although the circular electrodes are isotropic, the total detection system is anisotropic (the transfer function detection system was rotated with an angle $\theta = 25°$). Therefore, the design of an isotropic detection system requires the use of an isotropic spatial filter and an isotropic electrode shape.

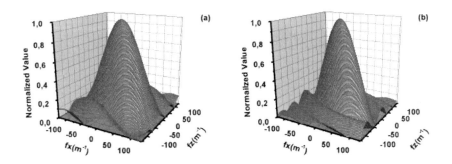

Figure 6. Absolute normalized value of the transfer functions related to the shape and dimensions of the: (a) rectangular (width and length of 7 *mm* and 15 *mm*, respectively) and (b) elliptical (short and large axes of 4,5 *mm* and 12 *mm*, respectively) electrodes with an angle of inclination of 25°.

Using 2D Spatial Filtering Techniques to Simulate Surface ... 209

When the surface SFAP signal is generated by an impulsive source, the 2D Fourier transform of the detected potential is obtained only by the multiplication the volume conductor transfer function and the detection system transfer function (Figure 8(a) and 8(b)). However, when the surface SFAP signal is generated by an analytical source, the 2D Fourier transform of the detected potential is the product of the Fourier transform of the current density source and the global transfer function. Figure 9(a) and 9(b) show the 2D inverse Fourier transform corresponding to the 2D Fourier transform of the signals shown in Figure 8(a) and Figure 8(b), respectively.

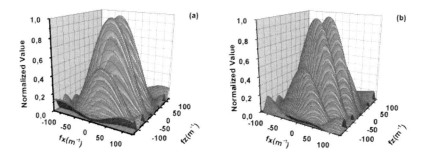

Figure 7. Absolute normalized value of the detection system transfer function constituted byTDD spatial filter with inter-electrode distances in the transversal and longitudinal directions of $dx = dz = 20$ mm and absolute normalized value of the detection system transfer function constituted by a grid of nine circular electrodes of radius of 15 mm with an angle of inclination of: (a) $\theta=0°$ and (b) $\theta=25°$.

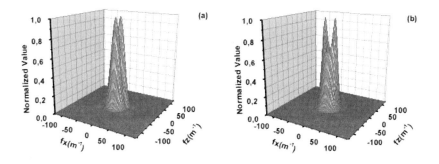

Figure 8. Absolute normalized value of the 2D Fourier transform of the potential detected on the skin surface (or the product of the global transfer function and the Fourier transform of the current density source) when the detection system is the contribution of a grid of nine circular electrodes and the spatial filters: (a) transversal double differential (TDD) and (b) Inverse binomial of order two (IB2).

The simulated surface SFAP signals generated by impulsive source are represented in Figure 10(a) and 10(b) and those generated by an analytical source are illustrated in Figure 10(c) and 10(d); the signals were filtered by the TDD and IB2 filters, respectively. The potential generated by an impulsive source is the result of the spatial phenomena (it is independent to the source parameters), while the potential generated by an analytical source is owing to the diameter and to the intracellular conductivity of the fibre.

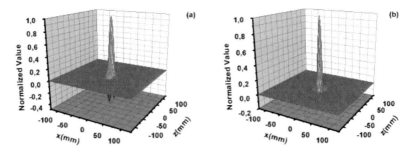

Figure 9. Absolute normalized value of the 2D inverse Fourier transform of the potential detected on the skin surface corresponding to the 2D Fourier transform of the results shown in figure 8.

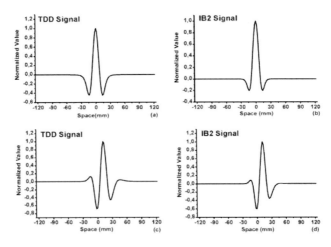

Figure 10. Absolute normalized value of the potential detected on the skin surface. These potentials are generated in a multilayer planar volume conductor by an impulsive ((a) and (b)) and analytical ((c) and (d)) sources. The spatial filters used for the configuration of the potential are: transversal double differential (TDD) in the case of (a) and (c) and the inverse binomial of order two (IB2) in the case of (b) and (d).

CONCLUSION

A model of surface SFAP signal generation and detection was presented and simulated by analytical approach. The volume conductor was modelled as a multilayer and anisotropic planar medium consisting of muscle, fat and skin tissue. The approach was based on the 2D Fourier transform properties. The results showed that the shape of the modelled SFAP strongly depends on the anatomical, physiological and detection system parameters. The findings also show that the SFAP generated by impulsive source does not have the same shape as that generated by an analytical source. In the case of the SFAP generated by an impulsive source, all the points of the signal are symmetrical with respect to the centre of the detection system. While in the case of SFAP generated by an analytical source, there is no exact symmetry. This result is due to the additional parameters of the source (diameter and the intracellular conductivity of the muscle fibre) in its analytical expression.

REFERENCES

[1]   Stegeman, D. F., Blok, J. H., Hermens, H. J., Roeleveld, K. (2000). Surface EMG models: properties and applications. *Journal of Electromyography & Kinesiology*, 10, 313–326.

[2]   Dimitrova, N. A., Dimitrov, G. V. (2003). Interpretation of EMG changes with fatigue: facts, pitfalls, and fallacies. *Journal of Electromyography & Kinesiology*, 13, 13-36.

[3]   Mesin, L., Cescon, C., Gazzoni, M., Merletti, R., Rainoldi, A. (2009). A bi-dimensional index for the selective assessment of myoelectric manifestations of peripheral and central muscle fatigue. *Journal of Electromyography & Kinesiology*, 19, 851–863.

[4]   Farina, D., Mesin, L., Martina, S., Merletti, R., (2004). Comparison of spatial filter selectivity in surface myoelectric signal detection – Influence of the volume conductor model. *Medical & Biological Engineering & Computing*, 42, 114-120.

[5] Stegeman, D. F., Blok, H. J., Hermens, J. H., Roeleveld, K. (2000). Surface EMG models: Properties and applications. *Journal of Electromyography & Kinesiology*, 10, 313–326.

[6] Merletti, R., Lo Conte, L., Avignone, E., Guglielminotti, P. (1999). Modelling of surface myoelectric signals – part I: model implementation. *IEEE Transactions on Biomedical Engineering*, 46, 810-820.

[7] Farina, D., Alberto, R. (1999). Compensation of the effect of subcutaneous tissue layers on surface EMG : a simulation study. *Medical Engineering & Physics*, 21, 487-496.

[8] Farina, D., Merletti, R. (2001). A novel approach for precise simulation of the EMG signal detected by surface electrodes. *IEEE Transactions on Biomedical Engineering*, 48, 637-646.

[9] Farina, D., Mesin, L., Martina, S., Merletti, R. (2004). A surface EMG generation model with multilayer cylindrical description of the volume conductor. *IEEE Transactions on Biomedical Engineering*, 51, 415-426.

[10] Mesin, L., Farina, D. (2004). Simulation of surface EMG signals generated by muscle tissues with inhomogeneity due to fiber pinnation. *IEEE Transactions on Biomedical Engineering*, 51, 1521-1529.

[11] Mesin, L., Farina, D. (2005). A model for surface EMG generation in volume conductors with spherical inhomogeneties. *IEEE Transactions on Biomedical Engineering*, 52, 1984-1993.

[12] Mesin, L., Farina, D. (2006). An analytical model for surface EMG generation in volume conductors with smooth conductivity variations. *IEEE Transactions on Biomedical Engineering*, 53, 773-779.

[13] Mesin, L. (2006). Simulation of surface EMG signals for a multi-layer volume conductor with triangular model of the muscle tissue. *IEEE Transactions on Biomedical Engineering*, 53, 2177-2184.

[14] Dimitrov, G. V., Dimitrova, N. A. (1998). Precise and fast calculation of the motor unit potentials detected by a point and rectangular plate electrode. *Medical Engineering & Physics*, 20, 374-381.

[15] Farina, D., Mesin, L., Martina, S. (2004). Advances in surface EMG signal simulation with analytical and numerical descriptions of the volume conductor. *Medical & Biological Engineering & Computing*, 42, 467-476.

Using 2D Spatial Filtering Techniques to Simulate Surface ...   213

[16] Lowery, M. M., Stoykov, N. S., Taflove, A., Kuiken, T. A. (2002). A multiple-layer finite-element model of the surface EMG signal. *IEEE Transactions on Biomedical Engineering*, 49, 446-454.

[17] Mesin, L., Joubert, M., Hanekom, T., Merletti, R., Farina, D. (2005). A Finite Element Model for Describing the Effect of Muscle Shortening on Surface EMG," *IEEE Transactions on Biomedical Engineering*, 53, 593-719.

[18] Blok, J. H., Stegeman, D. F., Van Oosterom, A. (2002). Three-layer volume conductor model and software package for applications in surface electromyography. *Annals Biomedical Engineering*, 30, 566-577.

[19] Messaoudi, N., Bekka, R. E., Ravier, P., Harba, R. (2017). Assessment of the non-Gaussianity and non-linearity levels of simulated sEMG signals on stationary segments. *Journal of Electromyography & Kinesiology*, 32, 70–82.

[20] Disselthorst-Klug, C., Silny, J., Rau, G. (1997). Improvement of spatial resolution in surface-EMG: A theoretical and experimental comparison of different spatial filters. *IEEE Transactions on Biomedical Engineering*, 44, 567-574.

[21] Farina, D., Cescon, C., Merletti, R., (2002). Influence of anatomical, physical, and detection-system parameters on surface EMG. *Biological Cybernitics*, 86, 445–456.

CONTENTS OF EARLIER VOLUMES

Advances in Engineering Research. Volume 29

Chapter 1   Performance Elements for
Future CMOS Technology
Gen Tsutsui, Heng Wu and Chun Wing Yeung

Chapter 2   Variable Porosity of Throughput and Tangential
Filtration in Biological and 3D Printed Systems
Alexander J. Werth

Chapter 3   Methane as an Alternative and Promising
Reducing Agent in Various Metallurgical
and Chemical Processes
H. Ale Ebrahim

Chapter 4   Layered-Type Transition Metal Disulfides
(TMDs) and their Composites as
Electrode Materials for Supercapacitors
S. V. Prabhakar Vattikuti

Chapter 5   Semi-Microscopic Consideration of
Minority Carrier Diffusion in Thin-SOI
and Wire pn-Junction Devices
Yasuhisa Omura

216   Contents of Earlier Volumes

Chapter 6   Active Voltage Contrast Imaging for Measuring
                 Electrical Potential Distribution Using Helium Ion
                 Microscopy
                 Chikako Sakai and Daisuke Fujita

Chapter 7   Two-Stage Study Program to Demonstrate the
                 Effectiveness of Injections in Capillary Masonry
                 Structures by Using the ESEM
                 Peter Körber

Chapter 8   Integer Asymmetric Error Control Codes for
                 Short-Range Optical Networks
                 Aleksandar Radonjic and Vladimir Vujicic

Advances in Engineering Research. Volume 28

Chapter 1   Special Features of a Functional Beam Splitter:
                 Diffraction Grating with Groove Bifurcation
                 Aleksandr Y. Bekshaev, Aleksandr I. Karamoch,
                 Anna M. Khoroshun, Oleksandr I. Ryazantsev
                 and Jan Masajada

Chapter 2   Adaptive Identification of Uncertain
                 Dynamic Systems
                 Nikolay N. Karabutov

Chapter 3   Wireless Sensor Networks
                 and Data Collection Methods
                 Minh T. Nguyen

Chapter 4   Antenna Array with Superdirectivity Properties
                 V. M. Koshevyy and A. A. Shevchenko

Contents of Earlier Volumes

Advances in Engineering Research. Volume 27

Chapter 1	Multi-Mode Dielectric Resonator Antennas *Ovidiu Gabriel Avădănei*
Chapter 2	Solution Processable Flexible Organic TFTs: Materials, Issues and Multidisciplinary Applications *Kshitij Bhargava and Vipul Singh*
Chapter 3	Six-Axle Locomotive Dynamics at Standard and Emergency Motion Conditions *Olga M. Markova, Helena N. Kovtun and Victor V. Maliy*
Chapter 4	Stochastic Digital Measurement Method and Its Application in Signal Processing *Aleksandar Radonjic, Platon Sovilj, Jelena Djordjevic-Kozarov and Vladimir Vujicic*
Chapter 5	Cornering Analysis of a Motorcycle Longitudinal Traction Control System *Andrea Bonci, Riccardo De Amicis, Sauro Longhi and Emanuele Lorenzoni*

Advances in Engineering Research. Volume 26

Chapter 1	Calcium Oxalate Crystallization in SDS Solutions *Yan Li, Xifeng Lu and Xiaodeng Yang*
Chapter 2	Development of Flash Ironmaking Technology *Hong Yong Sohn, Mohamed Elzohiery and De-Qiu Fan*
Chapter 3	Composite Materials Intended to Repair Mechanochemical Defects in Pipelines *H. F. Kudina, S. N. Bukharov and V. P. Sergienko*

218   *Contents of Earlier Volumes*

Chapter 4   Gibraltar Strait Barrier: Macro-Engineering
to Regulate Future Sea-Level Rise
Richard Brook Cathcart

Chapter 5   A Fast Parallel Seabed Image Texture
Classification Method Based on GPU
Reza Javidan and Seyyed Yahya Nabavi

Advances in Engineering Research. Volume 25

Chapter 1   Multifunctional Magnetic Sensors with
Applications for Traffic Management
Haijian Li, Honghui Dong and Limin Jia

Chapter 2   Vortex and Highly Turbulent Flows in the
Chemical Engineering: A Review of the
Specialized Software Products Cluster to
Perform Calculations for the Granulation Devices
Artem Artyukhov and Nadiia Artyukhova

Chapter 3   Piezoelectric Controlled Ceramics with
Metallic Implants: A New Alternative for the
Engineering Field
*Ernesto Suaste-Gómez, Daniel Hernández-Rivera
and Grissel Rodríguez-Roldán*

Chapter 4   Passive Magnetic Bearings
*Milad AlizadehTir, Fabrizio Marignetti
and Sorin Ioan Deaconu*

Chapter 5   A Universal Approach for the
Magnetic Force Calculation of Radial and
Axial Magnetic Bearings
*Ana N. Vučković, Dušan M. Vučković,
Mirjana T. Perić and Nebojša B. Raičević*

INDEX

A

abbe sine condition, 40, 42
action potential, v, viii, xi, 197, 198, 199, 200
ANFIS (Adaptive Neuro-Fuzzy Inference System), v, vii, viii, 47, 48, 50, 51, 52, 57, 90, 93
ANN (Artificial Neural Network), viii, 47, 48, 50, 187

C

CNC lathe, 144, 148, 153, 154, 156, 160, 161
CNC lathe machines, 144
computer numerical control (CNC), v, vii, x, 143, 144, 148, 153, 154, 156, 160, 161, 166
control, v, vii, viii, xi, 47, 48, 49, 50, 51, 54, 57, 60, 64, 65, 66, 67, 69, 70, 72, 73, 74, 81, 82, 85, 86, 88, 90, 91, 93, 94, 95, 96, 97, 98, 99, 100, 101, 102, 103, 104, 105, 106, 107, 108, 109, 120, 122, 134, 162, 166, 167, 168, 169, 171, 174, 175, 178,

180, 187, 188, 189, 190, 191, 192, 193, 194, 195, 216, 217
corrosion, v, ix, 111, 112, 113, 114, 115, 116, 117, 118, 119, 120, 121, 122, 123, 127, 128, 129, 130, 131, 132, 133, 134, 135, 136, 137, 138, 139, 140, 141
coupling efficiency, vii, viii, 1, 4, 10, 32, 39, 40, 41, 43, 44

D

degradation reinforced concrete structures, 112
detection system, xi, 197, 198, 199, 201, 203, 204, 205, 206, 208, 209, 211
DFIG (Doubly Fed Induction Generator), viii, 47, 48, 49, 50, 51, 57, 61, 62, 63, 64, 65, 67, 68, 70, 73, 74, 81, 82, 85, 92, 93, 94, 95, 96, 97, 98, 100, 101, 102, 103, 105, 106, 107, 108, 109, 189, 190
direction cosines, 3, 4, 9, 10, 21, 25, 26, 27, 28, 30, 36, 37, 42
divergence, 3, 4, 5, 6, 25, 26, 28, 30, 37, 42

Index

E

electromyography, 198, 211, 212, 213
eye-in-hand, vii, x, 143, 144, 145, 148, 153, 154, 155, 156, 160, 161

F

FL (Fuzzy Logic), v, 48, 51, 94, 96, 100, 101, 102, 103, 105, 106, 107, 108, 109, 167, 169, 170, 176, 177, 187, 189, 190
fuzzy, v, vii, viii, xi, 47, 48, 50, 52, 53, 54, 94, 95, 96, 100, 101, 102, 103, 104, 105, 106, 107, 108, 109, 162, 163, 167, 169, 170, 171, 176, 177, 178, 187, 189, 190, 191, 192, 194

G

gaussian beam, v, 1, 4, 28
generally astigmatic, vii, viii, 1, 2, 3, 4, 6, 7, 18, 21, 24, 25, 26, 27, 29, 36, 42

I

I/OLDC (Input/Output Linearizing & Decoupling Control), viii, 47, 48, 85, 86, 88, 90, 92, 93
insert inspection, vii, x, 143, 148, 150, 153, 154, 155, 156, 160, 161
intensity ellipse, 11, 18, 19, 20, 21, 22, 23, 25

M

machine vision, vii, x, 143, 144, 145, 148, 153, 154, 155, 160, 161
Memberships functions (MFs), xi, 167, 168
modeling, 43, 50, 94, 96, 97, 171, 188, 190, 194

O

online inspection, vii, x, 143, 144, 148, 153
optics, 2, 43, 44
optimization, viii, 1, 2, 3, 25, 26, 27, 28, 29, 30, 31, 32, 43, 44, 45, 188, 194
overlap integral, 39

P

parameter(s), viii, ix, x, 1, 2, 3, 6, 8, 9, 11, 12, 24, 25, 26, 27, 30, 49, 51, 52, 54, 55, 56, 70, 85, 86, 88, 90, 92, 93, 112, 115, 120, 129, 134, 164, 178, 179, 188, 199, 205, 206, 210, 211, 213
PI (Proportional Integral), ix, 48, 49, 50, 51, 57, 73, 85, 86, 87, 88, 89, 90, 91, 92, 93, 105, 169

Q

quadrotor, v, vii, xi, 167, 168, 169, 171, 172, 173, 174, 175, 179, 180, 187, 188, 189, 190, 191, 194, 195

R

ray, viii, 1, 2, 3, 4, 5, 6, 7, 8, 9, 10, 11, 12, 13, 14, 17, 18, 19, 20, 21, 24, 25, 26, 27, 28, 29, 30, 31, 32, 34, 35, 36, 37, 38, 40, 41, 42, 43, 44, 45
ray transfer matrix, 41
Rayleigh range, 27, 29, 32

S

Siding Mode Control (SMC), xi, 168, 169, 178, 180, 187

Index

221

simulation, vii, ix, x, xi, 48, 51, 53, 59, 61, 85, 86, 87, 89, 112, 138, 165, 168, 169, 179, 187, 194, 198, 212

spatial filter, viii, xi, 197, 198, 199, 201, 202, 203, 205, 206, 207, 208, 209, 210, 211, 213

spatial filtering, viii, xi, 197, 198, 199, 201

stability, 98, 169, 180, 187, 194

steel reinforcement, 112, 114, 115, 116, 117, 118, 123, 132, 135, 136

system, iv, v, vii, viii, x, xi, xii, 1, 2, 3, 4, 5, 6, 7, 8, 9, 12, 17, 21, 25, 29, 30, 31, 32, 35, 40, 41, 42, 47, 48, 50, 52, 57, 61, 65, 66, 68, 76, 82, 83, 85, 86, 88, 99, 121, 122, 131, 143, 148, 149, 151, 153, 154, 155, 156, 160, 161, 163, 165, 166, 167, 168, 169, 170, 171, 172, 174, 175, 176, 177, 187, 198, 199, 200, 208, 209, 211, 213, 217

T

tool inserts, 144

tracking, viii, xi, 48, 49, 51, 58, 59, 70, 88, 90, 93, 95, 97, 98, 104, 105, 106, 107, 167, 169, 180, 185, 186, 187

trajectory, xi, 167, 169, 179, 180, 187, 193

type-2 Fuzzy Logic Control (T2-FLC), vii, xi, 167, 168, 169, 175, 178, 180, 187

U

unmanned aerial vehicle (UAV), 168, 171, 173, 191, 194

V

volume conductor, xi, 197, 198, 199, 200, 203, 204, 206, 209, 210, 211, 212, 213

VSI (Voltage Source Inverter), viii, 47, 48, 95, 97, 102, 108

W

waist, 4, 5, 6, 7, 11, 18, 24, 25, 27, 29, 30, 35, 36, 37, 42

wavefront, 18, 20, 21, 22, 23

WECS (Wind Energy Conversion System), vii, viii, 47, 48, 49, 50, 51, 57, 81, 82, 94, 97, 98, 101, 102, 103, 106, 108, 189

Z

Zemax, 27, 29